NUMERICAL SOFTWARE TOOLS IN C

James Kempf

Software-Technology Laboratory,
Hewlett-Packard Company

PRENTICE-HALL, INC., Englewood Cliffs, NJ 07632

Library of Congress Cataloging-in-Publication Data

Kempf, James, (date)
  Numerical software tools in C.

  Includes bibliographies and index.
  1. Numerical analysis—Computer programs. 2. C
(Computer program language) I. Title.
QA297.K43   1987   519.4′028′5526   86-18740
ISBN 0-13-627274-6

PRENTICE-HALL SOFTWARE SERIES
Brian W. Kernighan, Advisor

Editorial/production supervision by Margaret Rizzi
Cover design by Lundgren Graphics, Ltd.
Manufacturing buyer: Ed O'Dougherty

UNIX is a registered trademark of AT&T Bell Laboratories
VAX is a trademark of Digital Equipment Corporation
HP-UX is a trademark of Hewlett-Packard, Inc.
PDP-11 is a trademark of Digital Equipment Corporation

Printed in the United States of America

10 9 8 7 6 5 4 3 2

ISBN 0-13-627274-6   025

Prentice-Hall International (UK) Limited, *London*
Prentice-Hall of Australia Pty. Limited, *Sydney*
Prentice-Hall Canada Inc., *Toronto*
Prentice-Hall Hispanoamericana, S.A., *Mexico*
Prentice-Hall of India Private Limited, *New Delhi*
Prentice-Hall of Japan, Inc., *Tokyo*
Prentice-Hall of Southeast Asia Pte. Ltd., *Singapore*
Editora Prentice-Hall do`Brasil, Ltda., *Rio de Janeiro*

CONTENTS

WHAT ARE NUMERICAL SOFTWARE TOOLS AND WHY PROGRAM THEM IN C?

Numerical programming is the neglected stepchild of modern software engineering. While the idea of software tools, pioneered by Brian Kernighan and Kenneth Plauger in their books *Software Tools* and *Software Tools in Pascal*, has transformed the art of writing systems software, numerical programming is still stuck in the "new program for every problem" stage. Most numerical analysis courses stress the mathematical and algorithmic side of numerical programming, and neglect the software engineering side completely. The result is numerical programmers that know a lot about numerical error, but very little about good software engineering practice. Numerical programs are often written as "throwaway software", useful for one problem only, after which they are discarded. Most good numerical programmers have had to learn through trial and error (or exposure to a programming course) that 1000 line subroutines and intricately nested GOTOs do not make software easy to maintain and change.

In the context of systems software, software tools are small programs which perform a limited number of tasks well. Their smallness generally facilitates testing and debugging. Such programs also have well-defined interfaces to each other and to the user. Careful thought in planning these interfaces makes software tools easy to use, both alone and together, so programmers tend to use them a lot. A collection of software tools might replace a large, amorphous system program with "features" that are sometimes hardly distinguishable from bugs.

To a certain extent, the increasing popularity of numerical libraries for doing such basic numerical tasks as Gaussian elimination or integration of differential equations has helped to improve the situation, at least with respect to algorithmic code. Because such libraries often take advantage of specific machine features and because they are generally written by professionals, results can sometimes be obtained using library subroutines with less CPU time and greater numerical accuracy than with programmer-defined numerical routines. Numerical library routines also speed production of numerical software, because they can be easily reused. However, as the everyday computation needs of scientists and engineers gradually migrate from the large mainframe, supported by a computer center staff, to a desktop micro, access to

FORTRAN libraries maintained by the computer center staff may gradually disappear, or become too expensive for certain applications. In addition, the "glue" code required to do input and output (including graphics), transform data from one form to another, and to generally organize data flow, must still be written. If this code can be written in a reusable manner, much time and programming effort can be saved when approaching a new problem.

This book is about how to apply the concept of software tools to numerical programming. The numerical programs and program modules described here are designed to work together, so that a collection of such programs becomes a superset of the typical FORTRAN callable library. The interface between the programs and the user is kept simple, to avoid having to learn a complex command set. Above all, the packaging of the numerical functions as discrete programs or program modules with well-defined interfaces simplifies the writing of numerical applications software. The coding effort is reduced, and much of the complexity of data structure definition is removed from numerical application programs, since it is hidden within the program and module building blocks. An application "program" is transformed into a string of several tools, working together, which can be assembled quickly to produce efficient results.

The central challenge of designing reusable numerical software is the existence of a problem-dependent model. Numerical problems, unlike systems programming problems, generally require that a real world system, like a bridge or a building, be modelled mathematically, and that this model be transformed into a computer code. The subroutines and functions embodying the model must then be linked with the appropriate analysis routines and embedded in code for organizing data flow to produce the final program. If the system text editor were constructed in this manner, a programmer would have to write a language- or application-dependent piece of code each time the editor were used and link it into the rest of the editor before the file could be edited. In this book, ways to reduce the amount of recoding necessary for different problems are discussed. While some numerical operations can be packaged as individual programs like a text editor, others require packaging as program modules which must be linked with the problem-dependent model code to produce the final program. An analogy can be drawn with the difference between a screwdriver and a socket wrench. A screwdriver can be used without modification on a wide variety of problems, while a socket wrench requires some simple preparation before it is ready for use.

A powerful programming environment provided by the operating system and a modern, structured programming language are crucial factors in the ability of a programmer to efficiently develop and use such numerical tools. Happily, the increasing acceptance of the UNIX system as the standard operating system for scientific and engineering programming is making such an environment available to more and more people. In a UNIX environment, the programmer has access to multiple processes, which allows several programs to run at once, and data pipes, which can act as channels between separately running programs. Facilities for maintaining collections of programming modules are available, and the operating system imposes no predefined constraints on data files, allowing the programmer more freedom in defining them. These features are by no means restricted to the UNIX system; however, on other operating systems they are often more difficult for the applications programmer to get at and use.

With languages, the situation is more complex. FORTRAN is and has been the standard numerical programming language since its introduction in the 1950's. Practically every engineering and science student has, at some point, taken a numerical analysis course in FOR-TRAN. The new ANSI standard, FORTRAN 77, even includes some features which more modern structured languages have, like character strings and structured IF-THEN-ELSE-ENDIF statements. FORTRAN compilers are widely available and often the local dialect contains additional features enhancing the ANSI standard. For this reason, Kernighan and Plauger used a FORTRAN dialect called Ratfor in their original *Software Tools* book.

Then why write this book about a language like C? Well, for one thing, despite the recent enhancements, FORTRAN still lacks some features which are essential in the development of numerical tools. For example, FORTRAN has only one kind of looping construct, the DO loop, and this loop has a fairly inflexible looping condition. A programmer who wants to encode a loop having a nonnumeric looping condition must either put a lot of thought into making something numeric that isn't or must use an IF-THEN-ELSE-GOTO-ENDIF loop, which may obscure the program's intent even more. FORTRAN also has only two types of data structures, scalars and arrays, neither of which is sufficient for organizing access to numerical data.

Pascal, another more widely known candidate, also has its drawbacks. While the structured control statements and the RECORD data structure give more flexibility to Pascal, the language as originally defined makes no provision for modular program development and data hiding. Pascal has no equivalent to the FORTRAN COMMON statement, which allows a programmer to limit what routines have access to globally known data. The lack of certain operators in Pascal, like the bitwise "and", forces recourse to assembly language at times. Pascal's rigid enforcement of strong typing also makes writing generic routines for particular algorithms difficult.

Until recently, an additional drawback of FORTRAN was the lack of compilers for microcomputers. Many numerical programmers were forced to use BASIC when working on microcomputers, since practically every microcomputer comes with a BASIC interpreter. Although some manufacturers have provided augmented BASICs with capabilities available in FORTRAN and Pascal, the ground level BASIC supplied with most microcomputers does not even have a decent subroutine construct. The GOSUB statement is no exception, since it makes no provision for passing parameters. Even today, most good quality FORTRAN compilers for microcomputers are relatively expensive.

C suffers from none of these problems. C has a full range of structured control flow constructs, plus a programmer definable type called the **struct** similar to the Pascal RECORD. C supports data hiding, in the form of **static** variables, whose scope is restricted to a particular function or file, but whose extent outlives the function invocation. C is also well integrated into the UNIX environment, and, even on non-UNIX microcomputer systems, reasonably priced C compilers are widely available.

On the other hand, there are certain aspects of C which make it difficult for beginners. The widespread use of pointers in C can be confusing. In C, pointers can be taken to practically anything, even functions. A beginner using a pointer to index through an array may accidentally index off the end of the array. The error might not be caught at runtime, as it would with FORTRAN and Pascal, until it's too late. This kind of problem can usually be avoided by good program design (as is the case in other languages as well), and, with the help of a symbolic debugger, can generally be traced down quickly if it does happen to occur.

The treatment of floating point variables in C is a more serious problem, particularly for numerical programming. High quality numerical programs require careful treatment of floating point, particularly with regard to precision, in order to achieve accurate results. C has contained both single precision (**float**) and double precision (**double**) floating point data types since its inception. But older C compilers often freely interconverted **float** and **double** variables during arithmetic operations, even if such interconversion was not strictly necessary. Even today, most C compilers convert single precision function parameters to double precision before a function call. This lack of control over floating point precision would seem like a fundamental obstacle to the acceptance of C for a successor to FORTRAN as the language of choice for numerical programming.

In part, the continued existence of automatic single to double conversion is part of C's legacy as a systems programming language. Since floating point is not particularly useful for systems programming, most compiler writers have not seen the need to clean up the automatic single to double precision conversion. However, as C grows into applications areas, like numerical programming, its childhood restrictions will be lifted. The recently drafted ANSI standard for C allows the declaration of the parameter type for functions in the calling function, so a compiler need not coerce single precision parameters to double precision. There are already some compilers available today that implement noncoercive floating point parameter passing by using an argument on the command line, which turns off the automatic conversion. While it may take several years before compilers implementing noncoercive floating point become widespread, many numerical tasks which today are done in BASIC or don't require stringent control of precision during function calls can probably be done without extra programming effort in C and will achieve acceptable accuracy. In addition, tasks with stricter precision requirements can be programmed, as long as the programmer compensates for automatic interconversion by rounding at the appropriate point in the code.

Perhaps the best advantage of C is that the UNIX operating system is written in it, so every machine running the UNIX system has a C compiler around somewhere. Using UNIX pipes, a programmer can break big programs down into smaller parts and run the programs simultaneously on a machine with limited memory, rather than using the more cumbersome overlay procedure which many other operating systems require. Since most modern computers make provisions for calling a routine written in one language from a program written in another, FORTRAN callable libraries can be used for the basic numerical analysis operations if desired, much as system routines are used for file i/o, trigonometric calculations, etc., in FORTRAN and Pascal.

The first chapter contains a quick introduction to C, stressing those aspects of the language which are most useful for producing numerical software. Language features not covered in Chapter 1 but used in later chapters are discussed when they are introduced. Some more obscure features of the language, like bit fields in **struct**s, are not discussed at all, since they are not needed to understand and write the numerical tools presented in the body of the book. Where they are needed, operating system functions from the UNIX system and C library functions are also introduced.

Chapter 2 presents three programs for performing linear algebraic operations on vector and matrix data. A common data exchange interface for tools using vectors and matrices is defined in Chapter 3. In Chapter 4, a simple program for plotting the results of numerical studies is discussed. The program depends on a graphics subroutine package containing appropriate routines for doing graphical operations on specific devices. Interfacing C to FORTRAN, Pascal, and BASIC is also discussed in Chapter 4.

In Chapter 5, tools for nonlinear optimization are presented. These tools build on the routines and tools previously developed. Finally, Chapter 6 presents numerical tools for integrating differential equations. In both Chapters 5 and 6, examples are given of real-world problems which can be solved by combining the numerical tools presented in those and previous chapters.

Although students having no background in programming could easily begin with this book, a familiarity with another programming language would certainly be helpful. Access to a C compiler is, of course, essential for programming the examples. Access to a machine running the UNIX system or a UNIX-like operating system, while not essential, will simplify some of the programming, since the UNIX system provides certain primitive operations which will come in handy. Though many of the example programs can be written for systems other than UNIX systems, some detective work may be necessary to ferret out the proper system functions. Features such as pipes and i/o redirection, which are relatively transparent on UNIX systems, are often more difficult to program on other operating systems.

Background material on algorithms is discussed when the algorithms are presented, and, where appropriate, every chapter begins with a short introduction to the analytic aspects of the numerical topic discussed in that chapter. Students should have prior exposure to calculus and a first course in differential equations for the material in Chapters 5 and 6, and simple matrix algebra for Chapters 2 and 4. No attempt is made to present mathematical proofs, nor are the derivations of the algorithms discussed in detail. Most algorithms are motivated heuristically, in order to leave space for discussion of their implementation. The reason is simple: while there are many fine numerical analysis books which discuss the mathematical side of numerical programming, very few treat the software engineering side.

I have resisted the temptation to introduce version specific code, so all the programming examples should be portable to every version of the UNIX system available. I have run them myself on HP-UX 3.0 and 4.0 (a System III derivative) on the HP9000/Series 500 and on Berkeley 4.1 and 4.2 on a VAX 11/780. In addition, the graphical filter in Chapter 4 was run on an HP9000/Series 200 workstation with a high performance display, running HP-UX 2.1, and on the HP Integral PC. Though I have not tried them on a Version 7 system, most of the system calls are generic enough that there should be no problem porting the code to Version 7 derivatives. I have avoided using the post-1978 enhancements (passing **struct**s to and returning them from functions, the **enum** and **void** types, etc.), although they are discussed briefly in Chapter 1, specifically to avoid problems with older compilers. About the only problem I encountered in moving between Berkeley 4.1/4.2 and HP-UX was the different name for the C library function which looks up the index of a character in a string (**strchr**() on HP-UX, **index**() on Berkeley 4.1/4.2).

Undergraduate students might find the book most helpful as a companion text for a first semester numerical analysis course or as the main text for a second semester course which emphasizes the software engineering side of numerical programming. Professional scientists or engineers who are not software engineers may find this book more useful for learning C than a book oriented toward systems programmers. Some estimates are that almost 50% of all software developed is numerical in nature, and, with the increase in the demand for computer-aided design and engineering applications, that number is not likely to get any smaller. The systems programming concepts presented herein will become increasingly important, as software architectures for applications become increasingly more complex.

I would like to acknowledge the help of Rod Junker, Martin Griss, Terrence Miller, Alex Gray, John Wilkes, and others at Hewlett-Packard for allowing access to and use of HP equipment in the production of the book and programming of the examples. The graphics filter in Chapter 4 was inspired by Chris Fraiser's graduate Computer Graphics class at the University of Arizona. Ian Birrell played the most important role of proofreader and judge. His diligence in reading through the entire manuscript several times contributed in no small way to the its quality, though, of course, the fault for any errors rests squarely on the shoulders of the author. Finally, I want to thank my wife Renate, for putting up with my spending many weekends and evenings over the last two years working. Her patience is to be admired.

1

C: THE BASICS

The C programming language is often called a "low level" language, because some of its operators correspond closely to assembly language instructions. For this reason, and because it compiles into fast executing code, it has become a favorite for system programmers. The popular UNIX operating system is written almost entirely in C. Despite this reputation, C has all the features of a powerful, truly general-purpose language. In some ways, it is more flexible than other languages which might be considered more suitable for numerical programming. Programmer-defined data types and recursive function calling are two features of C which are lacking in FORTRAN, for example.

The interface between the UNIX operating system and C is another plus, allowing numerical applications programmers to do system type programming easily. The enhancement in power and scope of the resulting application programs makes the UNIX system-C combination difficult to beat for producing general-purpose numerical software. Such programs, which are model independent so they can be reused from problem to problem, are the topic of this book. We'll call them numerical software tools, because, like the editors, compilers, and other system software tools which help programmers solve software engineering problems, numerical software tools help engineers and scientists solve numerical problems.

A C program is a collection of functions. Each function is a named subsection of the program. One function, called **main**(), is distinguished as the place where program execution starts. Execution begins with this main function and proceeds sequentially until the last statement in the **main**(), if all goes well.

In C, the body of each function is a separate scope whose local variables cannot be accessed from outside. Unlike standard Pascal, function definitions cannot be nested inside each other. Most C functions tend to be relatively short, although there are exceptions. If you find yourself defining a C function that runs over two or three printed pages, then it's probably time to re-examine the code and see if it can't be broken into smaller units. Short functions make C programs easier to read and understand, since each function does a limited number of things.

C is a typed language, and practically all variables and functions which return a value must be declared before use, though type checking is not strict, and can be circumvented if needed. Because the default type conversions are sensible, overriding type checking is rarely necessary. The C philosophy of typing is that types should serve as a guide and convenience to the programmer, rather than a willingly accepted straitjacket.

This first chapter gives an overview of the C language. Enough information on control structures, variable types, and the fundamental operators is presented so that we can get started talking about numerical software tools. Throughout the book, other features of the language, C library functions, and system functions for accessing features of the UNIX operating system are presented and discussed as needed.

The next section starts the chapter off with a look at a short, but not very useful, program.

1.1. A Quick Look at a Short C Program

Before going into a lot of detail on what kinds of operators C has, how its control structures work, etc., it might be helpful to get an overall picture of what a simple C program looks like. Here's a short program which multiplies two numbers together and writes out the result:

```
/*
******************************
mult-multiply two numbers together
    and write out the result.
******************************
*/

main()

{
  int x,y,z;

    x = 7;
    y = 6;
    z = x * y;

    printf("The answer is:%d.\n", z);

    exit(0);
}
```

```
/*
end of main
*/
```

When this program is compiled, linked, and executed from a terminal, the following line will appear:

The answer is:42.

Output in C is handled by a function (**printf()**) rather than through a statement in the language. As we'll see later, this is also true of input. Most versions of C come with an entire library of functions to do things like i/o, character string processing, sorting, and interfacing with the operating system, as well as the more traditional mathematical functions, like calculating sines and cosines.

printf() uses the quoted string, called a string constant, as a format conversion template. The other arguments, if any, are converted into character form according to the template and written out. We'll discuss the **printf()** family of output functions more in Chapter 2, but for now all we really have to know is that **%d** is a format conversion specification for printing out the integer **z** and (depending on the machine) the '\n' character causes a carriage return and line feed to be output after the text line, so that the next piece of text output with **printf()** starts at the beginning of the next line down on the terminal. The rest of the string is printed out exactly as is.

Every C program starts with three open files: standard input, standard output, and standard error. Unless some other arrangements are made, these files are usually associated with the screen and keyboard of the user's terminal. **printf()** writes to the standard output. Functions are also available which read from the standard input. Other files can be opened by function calls, and functions from the **printf()** family can be used to write to these files, about which more later. **printf()** and these functions are part of the C function library.

The opening and closing braces ({ }), enclose the definition of the function **main()**. A function definition usually begins with a series of variable definition statements. Practically all variables must be defined before use (the exception is discussed in Section 1.10), and there is no special type attached to particular beginning letters. Unlike FORTRAN, variables beginning with i through n are not guaranteed to be integer unless they are declared that way. Most versions of C distinguish between upper and lower case in variable and function definitions.

The program starts with a statement declaring the **main()** function. Every C program must have such a statement, indicating that the function definition to follow is the first function to execute when the program starts up. The **main()** is like any other function, except that the kinds of arguments which can be passed to it are restricted. For now, we won't need any arguments, though in Chapter 5 we'll return to the subject.

In the program we've declared three variables, **x**, **y**, and **z**, to be integers (**int**s). The integer constants **7** and **6** are assigned to **x** and **y** using the assignment operator, =. The multiplication operator, *, is used to multiply **x** and **y** together. These characteristics of C are not much different from any other algorithmic type language.

Every string of characters within the symbols /* and */ is ignored by the C compiler and treated as a comment. Comments can appear anywhere within a program. Statements themselves are not restricted to any particular part of a line as in FORTRAN, but can be placed so that they are pleasing to read and easy to follow. Indenting each new block level emphasizes the flow of the program, as long as the statements don't run off the right hand side of the page. Function calls and other statements which tend to crowd the right hand margin can be broken up into multiple lines, with individual pieces of the statement, like arguments with long names, placed on separate lines. Identifiers, like variable and function names, cannot be broken across lines, nor can they include embedded blanks.

C has no END statement telling when the program text is complete, but it does have a function which can be called anywhere within the program to cause execution to cease. This function is **exit()**. Although, strictly speaking, a call to **exit()** is not required at the end of **main()**, all programs in this book will make this call, since **exit()** communicates information to the operating system about the program's exit status. The integer argument is the channel for this information transfer. By convention, if the argument is zero, the program is considered to have successfully completed execution. If the argument is nonzero, then some fatal error occurred, and caused the program to abort prematurely. This distinction is not made automatically, however, and it is up to the programmer to call **exit()** with the proper argument depending on whether or not an error condition was encountered.

You'll notice that, during the previous discussion, all variable names, language keywords (like **int**) and function names were printed in boldface. Throughout the book, these objects, as well as names of files, commands, and executable programs are set off from the surrounding text, so we'll know whether we're talking about parts of a program or the programming environment. This makes it easier to talk about program text, since we can say "the **int**s **x** and **y**" rather than "the integer variables x and y". It also allows us to distinguish between program variables and variables used in the description of algorithms and other mathematical symbols.

If you want to get **mult** on your system, use a text editor to type in the source as **mult.c**. On the UNIX system, the C compiler is called **cc**. To compile and link your program, you need only type:

$ cc -o mult mult.c

and the program will be compiled, linked with standard library functions and put into an executable module called **mult**. The **$** is the UNIX command interpreter (shell) prompt.
If you had not included the argument **-o mult** on the command line, the executable module would have been given the default name **a.out**. Provided the program compiled and linked ok, you can start the program by typing:

$ mult

at your terminal. If you made no errors, the program should run.

1.1.1. Exercises

1. Try deleting parts of the program, or adding characters, and see what happens. Does the program ever get past the compilation stage before giving an error? Does it ever compile and link correctly, even though part has been deleted?

1.2. The while Loop

Suppose that, instead of just printing out the result of multiplying 7 times 6, we want a table of all multiples of 7 up to 7 times 12. We can modify our simple program to do this:

```
/*
*********************************
mult7-generate a table of multiples of 7,
        up to 7*12, using a while loop.
*********************************
*/

main()

{
  int i,x,y;

    i = 1;
    y = 7;

    while( i <= 12)
    {
      x = i * y;
      printf("7 times %d = %d.\n",i,x);
      i = i + 1;
    }

    exit(0);
}
/*
end of main
*/
```

mult7 is slightly more useful than **mult**, particularly if we happen to forget what the times table for 7 is.

The program also illustrates C's block structure. In C a sequence of simple statements enclosed in a matching pair of braces ({}) is called a block or compound statement. The program **mult7** above contains two blocks, one defining the function **main()** and another defining the body of the **while** statement. Examples of a simple statement are an assignment statement (like **i = 10;**) or a function call (like the call to **printf()** in **mult7**). Every simple statement must be terminated with a semicolon and a compound statement can be substituted wherever a simple statement occurs.

mult7 also illustrates one of the C looping constructs, the **while** loop. A **while** loop consists of the keyword **while**, followed by a logical expression in parentheses, after which comes a statement or compound statement that is under the control of the loop.

The syntax of the **while** loop is:

> **while**(expression)
>   statement;

The statement will not be executed at all if the expression in parentheses is not true when the **while** statement is executed. As long as the logical expression remains true, the statement will be executed. Usually some action is performed within the statement under control of the **while** loop, to cause the expression to become false. In the example above, the expression **i <= 12** becomes false when **i** reaches 13. Any variables appearing in the logical expression must be initialized before the **while** is executed.

1.2.1. Exercises

1. Change **mult7** so that it prints out the times tables for all numbers between 2 and 12 rather than just for 7. Start by putting another **while** loop around the outside of the one already there.

1.3. Relational Operators

C has no Boolean (TRUE-FALSE) type. Logical expressions evaluate to either zero, in which case they are false, or a nonzero result, in which case they are true. If relational operators like <= are used, the nonzero result is one, but there is nothing wrong with using another kind of expression which evaluates to something other than one.

Relational operators are used in logical expressions. The relational operators are >, >=, <, <=, ==, and !=. The last two mean "are equal" and "are not equal" respectively. The other relationals have their usual meaning, ie. the operator < means "is less than", etc. The "not" operator ! can be used to negate a logical expression. If an expression yields zero when evaluated, a ! before it will make it nonzero, and vice versa.

Here are some examples of how relational operators can be used:

> **1 < 2**   yields 1 or true
> **2 == 3**   yields 0 or false
> **!3**   yields 0 or false

Variables can be included in logical expressions, so:

      i <= j

yields one if **i** is less than or equal to **j**, zero otherwise.

One of the most common errors made when using the relationals is mistakenly typing = instead of ==. Such a mistake can lead to subtle bugs, for example:

```
while( i = 7)
{

    . . .

}
```

merrily sets **i** to 7 and executes the loop forever. The reason is that the assignment expression always evaluates to a nonzero value, hence the **while** conditional will never become zero, so the loop will never terminate.

This error is not flagged by the compiler because the assignment operator is like any other operator in C. It returns a value, namely the value of what is on the right hand side of the operator, making it possible to do multiple assignments on one line:

      x = y = z = 7;

Each individual assignment expression has the side effect of setting the variable to the left of it. While this property is useful, it requires that logical expressions be carefully checked to be sure that == and = are not confused.

1.4. The for Loop and Arithmetic Operators

One drawback of the **while** loop is that the initialization statement, looping condition, and incrementing statement are widely scattered about the code for the loop, rather than being concentrated at the top. This makes it difficult to see at a glance how the loop is supposed to behave.

If the initialization is done in another part of the program remote from the loop, or the looping condition is determined by some complex sequence of events, perhaps involving function calls, then the **while** loop makes more sense. If, however, the initialization and change in the looping condition are both done locally, then the **for** loop is a better choice.

The C **for** loop is outwardly similar to looping constructs in other languages, but there are subtle differences which make the **for** loop more powerful. Continuing with our example, we can modify **mult7** so that it uses a **for** loop rather than a **while**:

```
/*
*******************************
mult7-generate a table of multiples of 7,
    up to 7*12, using a for loop.
*******************************
*/

main()

{
  int i,x,y;

    for( i = 1, y = 7; i <= 12; i++)
    {
      x = i * y;
      printf("7 times %d = %d.\n",i,x);
    }

    exit(0);
}
/*
end of main
*/
```

The **for** loop is like a **while** loop with an initializing statement and an incrementing condition included at the top of the loop:

```
for(expression1; expression2; expression3)
    statement;
```

The first expression can be used to initialize variables for the loop. Expression2 is evaluated before each iteration (including the first) and the loop is exited if the result is zero. Expression3 can be used to increment variables and is executed after each iteration. If more than one initialization or looping expression are required, they must be separated by commas and the last one must be followed by a semicolon, as in the modified version of **mult7**. Any or all of the expressions can be omitted from the beginning of the loop. If all are omitted, the **for** loop is equivalent to a **while** loop for which the expression always evaluates nonzero; in other words, an infinite loop. Although anything can be put into the expressions at the beginning of a **for** loop, good programming practice suggests that only expressions associated with initializing and incrementing the loop be included. The statement controlled by the loop can also be compound.

The **for** loop in C is more useful than the Pascal FOR or the FORTRAN DO, since the looping condition need not be numeric, and can even be changed by the contents of the loop. For example, consider the following code fragment:

```
for( ; answer == YES; answer = affirm(input,answer) )
{
  input = ask();
  answer = change(input);
}
```

affirm() is a function which could change **answer**. The types of **answer** and **input** were deliberately left unspecified, and could be anything, as could the type of **YES**. The loop continues until **answer** is no longer equal to **YES**, regardless of what the types of **answer** and **YES** are.

1.4.1. Exercises

1. Repeat Exercise 1 from Section 1.2, except this time use a **for** loop. Keep the **while** loop as the inner loop.

2. Now repeat Exercise 1 from this section, using a **for** loop for both the inner and outer loops.

3. Have **mult7** print out a neat table of values, rather than simply printing out single lines. Note that you can use a **printf**() format string without the '\n' at the end, if you want to print out something without forcing a new line to start immediately afterwards.

1.5. Arithmetic Operators

The new version of **mult7** also illustrates a unique arithmetic operator in C, the autoincrement operator. A matching operator, --, can be used to autodecrement a variable. The autoincrement and autodecrement operators change the value of an integer variable by one. If **i** has value 2, then after the statement:

    i--;

is evaluated, **i** will have value 1. Similarly, if -- is replaced by ++, **i** will have value 3.

The autoincrement and autodecrement operators can be used in either prefix or postfix form. The prefix form puts the operator before the operand, as in **++i**, and means "increment the value of the variable before using it in the expression." The postfix form puts the operator after the operand (**i++**) and means "increment the value of the variable after using it."

As an example, suppose **i** has value 3. Then the effect of the expression:

    j = ++i;

is to autoincrement **i** to 4 and then assign **j** the value 4, while the expression:

> **j = i++;**

first assigns **j** the value 3, then autoincrements **i** to 4.

The postfix form of the autoincrement and autodecrement operators can often reduce the need for an additional assignment statement. Without using the autoincrement operator, the second example would have to be written:

> **j = i;**
> **i = i + 1;**

The autoincrement and autodecrement operators should be used with some care and forethought when combined with assignment. Inadvertently including the postfix form in an assignment expression where the prefix form should have been used, or the other way around, can lead to bugs. Most commonly, however, they are used for incrementing a loop variable and no assignment is taking place. In this case, either the prefix or postfix form will do.

Of course, C also has a full complement of the usual arithmetic operators. The multiplication operator (*) and the assignment operator (=) have already been mentioned. C has a total of five arithmetic operators: addition (+), subtraction (-), multiplication (*), division (/), and the modulus or remaindering operator (%), which in some other languages is a function. If **i** and **j** are integer variables, then

> **i % j**

gives the remainder when **i** is divided by **j**, if both **i** and **j** are positive. If either **i** or **j** or both are negative, the results are machine dependent. The modulus operator is only applicable to integers.

Expressions are evaluated left to right, except where parentheses force a particular subexpression to be evaluated first. Multiplication, division, and remaindering have precedence over addition and subtraction and will be evaluated first in a statement, unless the parenthesization changes the evaluation order. The **b + c** term in the first line of the example below will be evaluated before the division because it is in parentheses. There is no exponentiation operator in C. A function call must be used to do exponentiation. Also, integer division will truncate the fractional part and discard it.

The following C statements show some examples of these operators, along with the assignment operator:

> **q = a/(b + c);**
> **i = j % 2;**
> **s = r * (z = n - r);**

Note in the third example that the assignment can be included wherever an expression can. After the third line, **z** has the value **n - r** and **s** has value **r * (n - r)**. If an assignment expression is embedded in a larger arithmetic or logical expression, it should be enclosed in parentheses, since assignment has the lowest priority of all operators and will otherwise be done last.

Associative operators like + and * can be rearranged at the compiler's convenience, since the evaluation order isn't mathematically important. However, the order in which the terms of an expression are evaluated may be computationally important if the evaluation of a term changes the value of a variable appearing to the left of that term. This problem is discussed more thoroughly in Section 1.15, under a general discussion of side effects in C.

The freedom of the compiler to rearrange expressions which are theoretically associative can also change the result of a floating point addition, depending on the compiler used to compile the program or on how the expression is coded. A problem originates here because floating point arithmetic is only an approximation to real arithmetic. Due to the limited word size available for representing real numbers, when two floating point numbers are added, the result is truncated to fit within the machine's finite word length representation. The number of digits retained by the machine is the machine precision. If three floating point numbers of widely differing orders of magnitude are added together, the result can depend on the order in which the additions are done.

The following example might help clarify the issue. Suppose we have three floating point numbers:

$$a = 0.3456789$$
$$b = 0.4321098 \times 10^4$$
$$c = -b$$

If our machine has a precision of seven digits, then all digits beyond the first seven after the decimal point will be dropped during a floating point addition, so that the result will fit into the machine words used to represent floating point numbers.

Now suppose we want to add **a**, **b**, and **c** together. If we add **b** and **c** together first, the result is zero, since they exactly cancel. Adding **a** to zero gives **a**, or 0.3456789. If, however, we add **a** and **b** together first, the calculation becomes:

$$0.00003456789 \times 10^4 + 0.4321098 \times 10^4 = 0.4321443 \times 10^4$$

since only the first seven digits after the decimal point are kept. When we add **c**:

$$0.4321443 \times 10^4 - 0.4321098 \times 10^4 = 0.0000345 \times 10^4 = 0.3450000$$

the result is completely different than if we add the three together in the opposite order.

This difficulty is not exclusively confined to C, since it is true of floating point arithmetic in general. However, the fact that the compiler is free to rearrange expressions at random means that the results of a floating point calculation using a program compiled with one compiler might be completely different from those resulting if the program is compiled with another, or if the calculation had been programmed in a different way. If you suspect that the order of evaluation may be important for a particular calculation, intermediate calculations should be assigned to temporary variables and the temporary used in further calculations, forcing the order of evaluation to stay constant.

The kind of cancellation exhibited in the example can be avoided by carefully checking the number of significant digits in a result and truncating or rounding the value of a floating addition down to that value, if the order of magnitude indicates that precision may be a problem. In the example, only three digits were significant in the final result, and if the intermediate result generated by adding **a** to zero were truncated to three digits, the two would match. The extra computational effort necessary to check for precision problems on every floating point addition may not be necessary in the initial phases of an engineering or scientific investigation, but the final result of any numerical study is only as good as the accuracy of the calculation, and any numerical study should conclude with a thorough error analysis.

1.5.1. Exercises

1. Try programming the example of floating point addition using the C floating type **float** and numbers of varying precision. Does your machine truncate or round intermediate results to fit into a floating point word? How many digits of precision are retained? Note that, if your machine rounds intermediate results, the result of a calculation should be rounded instead of truncated (as in the example) to the proper precision.

1.6. The if-else Statement and Logical Operators

So far we have discussed looping, but we have no other way of changing the linear flow of a program. The **if-else** statement provides this capability.

Suppose we want to print out all the numbers between 1 and 100 whose squares are less than 100 or between 900 and 1000. We can do this by incorporating an **if-else** statement into the loop:

```
/*
**********************************
chksq-print out the number if its square
     is less than 100 or between 900
     and 1000.
**********************************
*/

main()

{
   int i,square;
```

```
                for( i = 1; i <= 100; i++)
                {
                  square = i * i;

                  if( square < 100)

                    printf("%d has square %d less than 100.\n",
                        i,square
                        );

                  else if( square >= 900 && square <= 1000)

                    printf("%d has square %d between 900 and 1000.\n",
                        i,square
                        );

                }

            exit(0);
        }
        /*
        end of main
        */
```

The **if-else** statement is used to determine which of the two conditions is satisfied by **i**. An **if-else** statement has the following syntax:

```
        if(expression)
            statement1;
        else
            statement2;
```

If evaluating the expression yields a nonzero result, statement1 is executed; otherwise, statement2 is executed. The statements can be either simple or compound and the **else** clause can be omitted from the **if** statement if no alternative action is desired. Chained **if-else** statements are often used to select one of a series of actions, based on whether one of a number of conditions is satisfied:

```
        if(condition1)
            statement1;
        else if(condition2)
            statement2;
          . . .
        else
            statementn;
```

Only one statement is executed, the statement for which the condition in the preceding **if** is true, or, if no condition is true, the statement after the final **else**.

Compound statements can be used within an **if-else** chain, exactly as within a **while** or **for** loop:

```
if( x > 0)
{
  sign = 1;
  printf("Setting sign positive...\n");
}
else if( x < 0)
{
  sign = -1;
  printf("Setting sign negative...\n");
}
else
{
  sign = 0;
  printf("x is zero...\n");
}
```

Here, an **if-else** chain is used to report the sign of **x** and record the result in the variable **sign**.

In nested **if** statements, care should be taken to ensure that the **else** portions are associated with the correct **if**. The intention of the following code fragment is to multiply **x** by **k** and record the result in **y**, if both **x** and **k** are positive. If **x** is nonpositive, then **y** should remain unchanged, but if **k** is nonpositive, an error message should be printed:

```
if( k > 0)
  if( x > 0)
    y = k * x;

else
  printf("Error:k is out of bounds.\n");
```

The indentation is meant to show that the **else** belongs with the first **if**; however, the compiler will interpret it in exactly the opposite sense. As a result, the error message will be printed when **x** is nonpositive instead of when **k** is. The final **else** clause is always associated with the last **if** lacking an **else**, as in most other languages.

1.6.1. Exercises

1.   Write a program, called **pair**, which finds all pairs of numbers between 1 and 100 that add up to 50 or 75.

1.7. Logical Operators

C has two logical operators which can be used to join subexpressions into more complex logical expressions for the conditions of **if**, **while**, and **for** statements. An expression containing the logical "and" operator, **&&**, evaluates to one only if all the logical subexpressions it connects do. C also has a logical "or" operator, **||**, which connects two logical expressions, and causes the entire expression to evaluate to one if either the left or the right subexpression do. The logical operators **&&** and **||** can be combined with the relationals in logical expressions.

As examples, consider:

(1 < 2) && (3 >= 4)

yields one, since both operands are true. Also

(2 == 3) || (4 > 1)

yields one since the right operand is true. But

(5 == 7) || (4 != 4)

yields zero since both operands are false. Also

(7 - 3 == 5) && (8 + 2 == 10)

evaluates to 0 since at least one (in this case, the left) operand is false.

This last example shows that logical, relational, and arithmetic operators can all be combined in one expression. Arithmetic operators are always evaluated first, eliminating the need for parentheses except to clarify a more complicated expression.

If the result of an arithmetic expression is either zero or nonzero, the arithmetic expression itself can be used as a logical expression and the relational operator == can be dispensed with entirely. For example, the following **if** conditional:

if(i % 2 == 0 || i % 3 == 0)
  . . .

evaluates to one if **i** is exactly divisible by 2 or 3. It could be rewritten without the == operator as follows:

if(!(i % 2) || !(i % 3))
  . . .

If **i** is divisible without remainder by either 2 or 3, then at least one of the expressions !(i % 2) or !(i % 3) evaluates to one, causing the code following the **if** to be executed. One disadvantage of not using the == is that the logic is slightly less easy to follow. The use of == provides a more visible way of specifying what the switching condition should be. Excessively complicated logical expressions using arithmetic operators should therefore be avoided.

C also supports ''short circuit'' evaluation of logical expressions. Logical expressions are always evaluated left to right, and evaluation stops as soon as the truth of the expression can be determined. If the expression involves the **&&** operator, the second operand will not be evaluated if the first is zero. Similarly, for the ‖ operator, the second operand will not be touched if the first is nonzero since the truth of the expression is already determined. Of course, if the correctness of further computations depends on the second operand being evaluated, short circuit evaluation can result in subtle bugs. This problem is similar to that which may occur in arithmetic expressions if the values of the operands to the left depend on the results of evaluating the operands to the right, and is discussed in Section 1.15, where the general problem of side effects is examined.

1.7.1. Exercises

1. Try writing some logical expressions that test short circuit evaluation. Because **printf**() is a function, it returns a value, namely the number of characters transmitted, or a negative number, if an output error occurred. One experiment you might try is putting **printf**() in the logical expressions to see what happens.

1.8. Other Control Statements

C has a number of other control statements that are occasionally useful, but the **if-else**, **while**, and **for** statements account for more than 90% of the control code in most C programs. Useful C programs can be written without any of the other control statements, though extra variables may be necessary to implement the control logic. At points in the book where these statements are used, they are discussed.

One control statement that deserves mention (which, incidentally, will NOT be used anywhere else in the book) is the venerable **goto**. C, like many other languages, has an unconditional **goto** statement. In the following example, a **goto** statement is used to cause a jump to error handling logic if the nonnegative integer variable **i** contains an invalid value:

```
if( i < 0)
   goto error;

error:
   printf("Error:i must be zero or positive.\n");
   exit(1);
```

The syntax of the C **goto** is simple:

      goto label;

      . . .

      label: statement;

The label must have the same form as a valid C variable name, and is restricted to being within the same function as the **goto** statement. After the **goto** is executed, control transfers unconditionally to the statement after the label.

C's control statements make **goto**s practically unneeded. In most cases, introducing control variables or rearranging the logic can get rid of a **goto** that seems to be required. The result is usually a program which is considerably more readable, and easier to maintain. Excessive use of unconditional branching can often lead to a tangled mass of control logic which is difficult or impossible to follow. In FORTRAN, unconditional branching is practically impossible to avoid because the lack of flexible looping structures make loops using GOTOs necessary. C's set of powerful looping constructs eliminates the need for looping **goto**s.

1.9. A Useful Example: Finding Prime Numbers

Before continuing with our discussion on C, let's pause for a moment and consider an example program, called **prime**, which is more useful than the previous ones. **prime** contains all the control constructs and a number of the arithmetic and logical operators that we have already discussed.

A positive integer i is said to be prime if i is greater than 1 and the only two positive numbers which exactly divide i are 1 and i itself. The first few primes are easy to find; they are 2, 3, 5, 7, 11, 13, 17, 19 ... Higher primes become progressively more difficult to find and must be calculated.

prime prints out all the prime numbers less than 100:

```
/*
**************************************
prime-print out all the primes less than 100.
**************************************
*/

main()

{

  int i,j,max,min;

    max = 100;
    min = 2;

    for( i = min; i <= max; i++)
    {

      j = 2;
/*
 iterate on j, as long as j doesn't exactly
    divide i
*/

      while( i % j++ != 0);

/*
 now check if we've found a prime
*/

      if( j == i)
        printf("%d is a prime.\n",i);

    }

    exit(0);

}
/*
end of main
*/
```

The lower and upper limits, **min** and **max**, between which primes are sought could be changed to any two integers. The outer **for** loop iterates over all the integers between **min** and

max. At the worst, the inner **while** loop iterates over all the integers between 2 and **i**, though it could break off earlier if **i** is not a prime. For this reason, the inner **while** loop is kept short, to reduce the amount of computation each time around. The autoincrement operator is used in the postfix position in the **while** loop so the value of **j** is used in the modulus operation before **j** is incremented. The body of the **while** loop is an empty statement, denoted by the semicolon immediately after the **while**. The value of **i** is only printed if the **while** loop terminates when **j** equals **i**, since that means that **i** is prime.

1.9.1. Exercises

1. An integer j, is said to be relatively prime with respect to another integer i if the largest integer which divides both exactly (the greatest common divisor) is 1. For example, 6 and 25 are relatively prime because 1 is their greatest common divisor, while 6 and 8 are not, because their greatest common divisor is 2. Write a program, **rprime**, which prints out all the relative primes for a particular integer between an upper and lower limit.

1.10. More on Identifiers, Constants, and Typing

In the previous sections, we've informally used variable names, declarations, and constants without discussing them in detail. String constants were briefly mentioned in connection with **printf()**, and integer constants were sprinkled throughout the examples. **int** variables were declared in each of the programs. In this section, we'll discuss the formal rules for C identifiers, what legal C constants are, and some other types besides **int**.

Names of programmer-defined variables and functions are called identifiers. Identifiers in C can be made up of the lower and upper case letters, digits, and the underscore (_) character. An identifier must begin with a letter or an underscore and only the first eight characters are significant (though this varies from compiler to compiler).

For example:

> **x**
> **new_number**
> **y22**
> **_last**
> **NOT_THIS_ONE**

are all examples of valid identifiers. Note that in most C implementations, **x** and **X** are two different identifiers because their case differs. No embedded blanks are allowed in an identifier. The following are not valid identifiers in C:

> **last car**   contains an embedded blank
> **1y**   begins with a number

while the two identifiers:

thisvariableisverylong
thisvariableisverylongtoo

are the same for many C compilers, since they are alike in their first eight characters.

There are some identifiers which cannot be used for any programmer-defined objects. These keywords are part of the C language. Table 1.1 contains a list of the C keywords. **ada**, **asm**, **fortran**, and **pascal** are used for embedding other languages within C, while **entry** was originally reserved but was never used. The keywords **void, enum, entry, asm, pascal, ada,** and **fortran** may not be implemented on some compilers.

int	char	float	double
long	short	unsigned	auto
extern	static	register	struct
union	typedef	goto	return
sizeof	break	continue	if
else	for	do	while
switch	case	default	void
entry	asm	pascal	ada
fortran	enum		

Table 1.1. C Keywords.

A C integer constant is simply a sequence of digits between 0 and 9. If the constant does not begin with a zero, the compiler interprets it as a decimal (base 10) integer. Valid decimal constants are **10**, **5280**, and **10000**, while **012** and **1LN** are not. If the integer constant begins with a zero, as in **012**, the compiler interprets it as an octal (base 8) integer. Some valid octal constants are **012** (decimal 10), **005** (decimal 5), and **000** (decimal 0). An integer constant can be as large as the machine word size allows.

Floating constants consist of a whole part, a decimal point, a fractional part, followed by an **e** or **E**, and a signed integer exponent. The whole part and the fractional part are integer constants. Either the **e** or **E** (with or without the signed integer exponent), or the decimal point must be there for the number to be a valid floating constant. Similarly, either the fractional or whole part are required. All floating constants are stored in double precision.

Some floating constants which illustrate the range of possibilities are:

2.5	contains whole part and fraction but no exponent
2.	contains only whole part
.5	contains only fractional part
715.632e	contains whole part, fractional part, and **e** but no integer exponent
43.0E-22	contains whole part, fractional part, and integer exponent with sign
34e5	contains no decimal point

In practice, constants like the fourth example above should probably be avoided, since the value of the exponent, while interpreted by the compiler to be zero, appears ambiguous.

Some examples which are not floating constants are:

e5	contains no whole or fractional part
.e+3	contains no whole or fractional part
1234	integer constant
x22	C identifier
1.34e-2.5	exponent must be an integer

A character constant is a single character enclosed in single quotes, like **'a'** or **'?'**. The numerical value of a character constant is the value of the character in the machine's character set. Most machines use the ASCII set, based on representing characters as 8-bit (one byte) integers. In the ASCII character set, all the letters of one case are in increasing sequence, as are the numbers. There is, however, no sequential relationship between the letters of one case and those of another, nor between the letters and the numbers. The numerical value of the letter **'A'** is less than that of **'Z'** and similarly for **'a'** through **'z'** and **'0'** through **'9'**, though **'A'** is not greater than **'a'**. Similarly, the value for the character **'7'** is not the same as the integer constant **7**, though the value of **'7'-'0'** is 7, since the numerical values of the number characters do occur in increasing sequence. Other, less commonly used character sets are EBCDIC and Display Code. We'll assume throughout the book that the character set we're working with is ASCII.

String constants are sequences of characters surrounded by double quotes, ". In the next section, we'll talk about arrays in more detail, however, strings are represented internally as arrays of characters, on the very end of which a null byte (i.e. **'\000'**) has been inserted. **"Susanne"** and **"that bird"** are examples of legal C string constants, while **"Who** is not, since the closing double quotes are missing. **'this dog'** is also incorrect, since single quotes surround character, not string, constants. However, **"a"** is a perfectly good string constant, since a single character can be considered a string. The string constant **""** is also legal, and is sometimes called the null string, since it consists of just the null byte which terminates the string. Null strings are to string constants what zero is to integers.

A limit, varying from compiler to compiler, is enforced on the maximum size of string constants. If a string constant must run over the end of a line, a backslash (\) can be used to tell the compiler to ignore the end of the line and continue interpreting the characters on the next line as part of the string. As a practical matter, multiline string constants tend to be difficult to read and should be avoided if possible.

There are certain characters which cannot be printed by a terminal or line printer but which are useful to have, since they tell the output device to do something special. One example is the newline character, which tells the terminal to return the cursor to the beginning of the line and move down one line, so typing can begin on a new line. We've mentioned the newline character in the previous sections. It is represented by the C character constant **'\n'**. The backslash (\) is an escape convention which tells the C compiler that the following n isn't really an n but rather, when printed on a device which understands what to do with it, causes further text output to be printed on a new line. A list of the nonprinting characters and other characters to which the escape convention can be applied is shown in Table 1.2.

newline	**\n**
tab	**\t**
backspace	**\b**
carriage return	**\r**
form feed	**\f**
backslash	****
single quote	**\'**
double quote	**\''**
bit pattern	*****xxx*

Table 1.2. Escape Sequences.

Note particularly that '\'' is the way to get a character constant that is a single quote and '\''' gets a string constant that is a double quote. The backslash character itself can be represented as '\\'. Any character not on the list in Table 1.2, when preceded by a backslash, is the same as the character itself, so '\j' and 'j' are the same character constant. Finally, if a one-, two-, or three-digit integer constant within single quotes is preceded by a backslash, the constant is interpreted as a character. For example, the octal character constant '\007' is the unprintable ASCII character formed on most keyboards by pressing the 'CONTROL' and 'G' key at the same time. The CONTROL G is sometimes called BEL, since, when sent to the terminal, it causes the terminal to beep.

The **int** type was introduced in the programming examples. An integer is of type **int** and usually has a machine representation equivalent to one machine word. C has three other predefined types with which we'll be concerned in this book: **char**, **float**, and **double**. Identifiers of declared type **char** are eight bit positive integers which are used to represent characters in the machine character set. A **float** is a single precision floating point number while a **double** is a double precision floating point number. As previously mentioned, practically all identifiers (variable and function names) must be declared before use. The exception is identifiers of type **int**. Any variable or function name which is undeclared is assumed by the compiler to have type **int**. However, even variables and functions of type **int** should be declared, since the declarations serve as a kind of self-documentation and can sometimes prevent subtle bugs from cropping up if the code happens to change.

The following are some declarations of identifiers:

```
int p,q;
float ix,iy,iz;
double f();
```

The semicolons make the declarations valid C statements. The last entry declares the identifier **f** to be a function returning a double precision floating point number.

C is rather loosely typed and certain type conversions occur automatically. The conversions are similar to those in other languages. Before an expression is evaluated, all operands with type **char** are converted to **int**s. After that, if the expression still contains mixed types, a hierarchy of type conversion is imposed on the expression. Variables of type **double** are highest in the hierarchy, followed by variables of type **float**, with variables of type **int** on the bottom. If an expression contains variables in two or more levels of the hierarchy, the variables at the lower levels are promoted to the highest level before the expression is evaluated. An expression containing an **int** and a **double** will have type **double**, similarly, an expression containing a **float** and an **int** will have type **float**.

Early C compilers converted all **float** identifiers to **double** before doing any arithmetic and before passing **float** parameters to functions, even if all the operands in the arithmetic expression or function call were **float** and the conversion was technically unnecessary. This language "feature" was an artifact of the hardware on which the language was originally developed, the PDP 11 series, and not the result of any technical or scientific considerations in language design. Many currently available low-cost microcomputer C compilers still do **float** to **double** conversion before all operations, while others only do **float** to **double** conversion when passing **float** function parameters. Some higher quality C compilers have begun to offer an option which turns off automatic **float** to **double** conversion, but most of the supplied mathematical functions (like **sin**() and **cos**()) only accept **double** parameters and return **double** results. When designing high-quality numerical software, more control over the precision of floating point numbers is often necessary, and leaving this critical function up to the compiler is sometimes unacceptable. As C becomes more popular and standards begin to evolve, it seems likely that unnecessary automatic **float** to **double** conversion will be dropped and function libraries for both **float** and **double** will begin to be offered. In the meantime, the safest course (and the one adopted in this book) is to program all real arithmetic in **double**. If memory limitations dictate, however, **double** identifiers can be programmed as **float**.

Conversions can also occur during assignment. The conversion works as expected; that is, the type of the expression is the same as that of the variable on the left hand side of the =. So, for example, in the following expression:

```
float f;
int i;

i = f;
```

the value of **i** after the assignment is exactly the whole part of **f**.

In the case of a **char** to **int** conversion, if the ASCII value of the **char** is positive, the upper bytes of the machine word for the **int** are padded with zeros. If the ASCII value of the **char** is negative, however, the upper bytes could either be padded with zeros or ones, depending on the machine. It is possible to have a negative **char** variable, but C guarantees that the usual characters in whatever machine character set is used will have nonnegative values. The effect is that assigning a **char** to an **int** and then assigning the **int** back to a **char** will leave the **char** unchanged.

Two more recent additions to C are the **void** and **enum** types. **void** has no values and no operators, and is primarily used to indicate that a function returns no value:

> **double x[10];**
> **void read_vector(x);**
>
>   **read_vector(x);**

The **enum** type is similar to Pascal enumeration type:

> **enum color {red,green,blue};**
>
> **color dot,box;**

An enumeration is a set of values represented by identifiers, like the colors in the example above. The declaration creates a new enumeration type, called **color**, having three values: **red**, **green**, and **blue**. The variables **dot** and **box** can only take on the values **red**, **green**, and **blue**.

Many C compilers do not implement either the **void** type or the **enum** type, so neither will be used in this book. **void** is primarily an aid to program readability, since, even if a function returns a value, C allows it to be called without assigning the return value to a variable. The **enum** type is also, and, in addition, compilers which implement **enum** often do so inconsistently. Constants defined using the C preprocessor (discussed in Section 1.16) generally serve about the same function and are considerably more flexible, and more portable too.

In addition to **int**s, **char**s, **double**s, and **float**s, C contains a number of other types. One of these, the programmer-defined data type, is discussed in Chapter 2, while some others are briefly mentioned in the exercises accompanying the last section in Chapter 4. The rest aren't very useful for numerical programming, and therefore won't be discussed further in this book. If you're interested in learning more about C types, consult one of the books listed in the reference section at the end of the chapter.

1.10.1. Exercises

1. Modify **mult7** to use **float**s and **double**s instead of **int**s . The format specification for printing out **float**s and **double**s in **printf**() is **%g**.

2. Write a program called **chnum** which prints out the numerical values of some characters. The **%d** format specification can be used for printing out the numerical character representation, while the **%c** specification can be used to print out characters as characters.

3. Write a short program which tests what happens when a negative integer, or integer outside the range of ASCII characters, is converted to a character. Have the integer converted to a character then back to an integer and print out the result. How does your machine handle integer to character conversion for these two cases?

1.11. Arrays and Pointers

Arrays in C are similar to arrays in other languages, with two important differences:

1.   The first entry for any array in C has index zero instead of one.

2.   Multidimensional arrays in C are stored by row, as in Pascal, instead of by column, as in FORTRAN. When array indices are calculated the last index varies the quickest, rather than the first.

In C, arrays must be declared before use to indicate to the compiler that it should allocate storage. The following code fragment illustrates how arrays are declared:

```
int q[7],p[8][8];
double x[20];
```

The first elements in each of the above are **q[0]**, **p[0][0]**, and **x[0]**, respectively, and the last elements are **q[6]**, **p[7][7]**, and **x[19]**. The index of the last element is always one less than the declared size of the array.

The zero-based indexing system takes a bit of getting used to, since most other languages either allow the programmer to specify what the index base is or start with one. FORTRAN programmers accustomed to arrays starting with one might simply elect to ignore the zero-based indexing scheme in C and start all computations with the second row, whose index is one. The danger with this strategy is that the last index in the array is not equivalent to the declared size of the array, as in FORTRAN, but to that size less one. The arrays must therefore be declared one larger than actually needed, which could be confusing while programming. For these reasons, we'll follow the zero-based indexing scheme in the remainder of the book.

Character arrays deserve some special discussion. As mentioned previously, string constants are stored as arrays of characters with a null byte appended. String variables can also be used, and are declared as arrays of characters:

```
char string[20];
```

When declaring character string variables, be sure to declare enough space for the null byte. The above declaration, for example, will accommodate a string of nineteen characters, maximum, in the array elements **string[0]** through **string[18]**. The last position, **string[19]**, is filled by the null byte.

C does not allow block assignment of arrays. Copies of strings from one string variable to another must be done element by element, as must copies of numeric arrays. This is not as serious a programming problem as it might seem, since the library of functions supplied with the C compiler contains several functions to do string manipulations.

Multidimensional arrays are declared like the two-dimensional integer array **p[][]** in the example at the beginning of this section. The largest multidimensional array of practical interest is the two-dimensional array, which is a convenient data structure to represent a matrix.

Although larger arrays can be declared, the meaning of the subscripts often becomes confusing. In most cases, a programmer-defined data structure is easier for the next person who looks at the program (which may be you) to understand, and organizes the information more coherently. Programmer-defined data structures are discussed in Chapter 2.

A two-dimensional array in C can also be thought of as a one-dimensional array with each element itself being a one dimensional array. Because C arrays are indexed with their last subscript varying fastest, the most convenient way to think of a two-dimensional array is as a vector of rows in a matrix.

The actual location in memory of the i,jth element in the two-dimensional array **p[][]** is calculated as:

$$(\text{address of } \mathbf{p[0][0]}) + i * 8 + j$$

C provides an even better way of expressing this calculation. The identifier declared as the array name in C can be used as the address of the first memory location in the array. The index calculation can be written more simply as:

$$\mathbf{p + i * 8 + j}$$

When the array name is used in this way, it is called a pointer. A pointer can be thought of as the address in memory of some data. A pointer "points" to the storage cell in the computer's memory where that data is stored. In order to get at the data, the address must be followed and the data fetched, a process called dereferencing.

For example, if an integer is declared as:

int q;

then a pointer can be taken to **q** by prefixing **q** with the ampersand, or address, operator:

&q

The result of applying the **&** operator to an identifier is to produce the memory address of the identifier.

Using an assignment statement, the pointer can be stored:

p = &q;

The variable **p** must also be declared, and the declaration looks like this:

int *p;

The asterisk indicates that **p** is a pointer. The asterisk is also used as an operator to get at the contents of the location pointed at by the pointer, the operation defined above as dereferencing. If **r** is another declared **int**, then the effect of:

r = *p;

is exactly the same as:

r = q;

The contents of the storage location called **q** will be transferred to the location called **r**. The location **p** will contain the address of the location **q**.

Although pointers give programmers great freedom in manipulating how and where things are stored in memory, they can also be confusing and downright dangerous. Attempting to access an array position beyond the memory allocated in the array definition statement is a commonly occurring bug in any language. With pointers, this kind of problem becomes even more serious. The effects of using a pointer that points to an address which the programmer didn't intend to access may not show up until after the program has progressed beyond where the illegal pointer reference was made. This can make the illegal reference very difficult to find.

Furthermore, since the amount of memory allocated to arrays must be declared beforehand, the run time support for a language can often check whether an array reference is within bounds and can stop the program if not. With some kinds of pointers this is impossible, since no declaration on the amount of memory to reserve is made beforehand. If the program is running on a microcomputer with little or no memory protection, a wild pointer can result in an operating system crash.

Another common pointer problem is an attempt to dereference a pointer which was not initialized or which has been initialized to zero. The C compiler doesn't initialize pointers, so all pointers should be initialized to zero in the program before they are used. An attempt to dereference a null or uninitialized pointer may or may not cause the program to crash, but should be avoided in any case. Initializing pointers to zero makes it easy to catch a pointer which has not yet been assigned a valid address.

By making use of the fact that the array name can be used as a pointer to the first memory location in the array, pointers can also be used to access elements of an array. The value of the array element **x[20]**, for example, can also be accessed as ***(x + 20)**, since **x + 20** is a pointer to the 20th element of the array and the ***** operator dereferences it. The reason why C uses a zero based indexing system for arrays should now be obvious. **x + 0** yields the same address as **x**. In general, the *i*th element of an array can be accessed either by adding *i* to the pointer in the array name variable, then using the ***** operator to dereference the pointer, or by using the standard square bracket notation and the array index.

The cautions expressed above with regard to wild pointers are also applicable to using pointers in arithmetic expressions. Pointers are not integers and trying to do arithmetic with them as if they were can lead to nonsense results. In general, adding an **int** to a pointer pointing into an array will result in a correct address calculation, provided the calculated address doesn't go beyond the end of the array. Indexing the array position **x[20]** is an example.

Subtracting an **int** from an array address is also allowed. The compiler assures that the correct number of bytes for the data type involved is added to or subtracted from the pointer as an offset.

Similarly, if **p** and **q** both point to array positions at different points in the same array, then **p - q** yields an **int** representing the number of array elements between them. If, however, **p** and **q** point into different arrays, or into arrays of different types, then subtracting the two is like subtracting apples from oranges. Addition or subtraction of an **int** from a pointer and subtraction of two pointers which point into the same array are the only two arithmetic operations which should be attempted on pointers.

1.11.1. Exercises

1. Rewrite **mult7** to use pointers to **int**s, **float**s, and **double**s.

2. Write a program, called **strfill**, which fills a character array with a string, then prints it out character by character and also as a string. The entire string can be printed by using **printf**() with the **%s** format conversion.

3. Modify **strfill** to use pointer arithmetic rather than indexing to move through the array, and pointer dereferencing to get at a character. Add a section of code which calculates the number of elements between two array positions by subtracting pointers.

4. Try writing a program which dereferences:

    i.   an uninitialized pointer,

    ii. a pointer whose value is zero,

    iii. a pointer which points outside an array.

    What happens? If you are running on a microcomputer, you will want keep an operating system diskette handy for this one.

1.12. Functions in C

In C, the main program is just a function with the special name **main**(), as the examples at the beginning of the chapter illustrated. The only difference between **main**() and other functions is that the parameters with which **main**() can be declared are restricted. The examples in previous sections showed **main**() without parameters, but in Chapter 5, the special parameters for **main**() are discussed.

Other functions can be defined with any mixture of parameters as needed, though the compiler may put a limit on the total number of parameters a function may have. All executable code (as opposed to global data declarations and preprocessor statements, discussed later in this chapter) must be within function definitions. The formal outline for defining a function in C is:

> type <function name> (parameter list)
>   parameter list declarations
> {
>   local variable declarations
>
>   function body
>
>   **return**(value returned);
> }

If the function does not return a value, then the **return** statement need not include a return value. **return** statements can be placed anywhere within the function body.

The <function name> must be a valid C identifier. The type is optional. If no type is declared, the compiler will assume that the function returns an integer. In practice, functions which return **int**s should also be declared, since the declaration serves as a form of documentation for the next person to look at the program. If the return value is never used, the type can be omitted and the **return** statement can simply be written:

> **return;**

In addition, if the compiler supports the **void** type, a function which doesn't return a value can be declared **void**.

The parameter list declarations declare the type of the parameters in the parameter list. If the function is not passed any parameters, the declaration can be omitted, although the parentheses in the header cannot. A simple example of a function definition is:

```
/*
*************************
abs-return the absolute value
    of the argument.
*************************
*/

int abs(i)

  int i;

{
  if( i >= 0)
    return(i);

  else
    return(-i);
}
/*
end of abs
*/
```

which is an implementation of the absolute value function. If **i** is positive or zero, the value of **i** is returned; otherwise, the function returns **-i**.

Variables declared within the braces that surround the body of a function are called local variables. Local variables are only directly accessible within that function, and are only defined when the function is called. These local variables disappear when the function returns. For example, in the following function:

```
/*
*************************
max-return the maximum of
    the two arguments.
*************************
*/

int max(i,j)

  int i,j;

{
  int k;

  if( i >= j)
    k = i;
```

```
        else
          k = j;

        return(k);
}
/*
end of max
*/
```

k is undefined before **max()** is called and after it returns. **max()** implements the maximum function, returning the larger of its two arguments.

A function call in C can be part of an assignment statement, part of an expression, or can stand alone as a statement itself. If the return value of the function must be saved after the function is called, then it can be assigned to a variable in the calling routine. If the return value of the function is not needed, the function can be called without an assignment statement. A function call can be substituted wherever a variable of the function's return type occurs. Some examples of legal function calls to the functions defined above are:

```
q = abs(q);

if( max(x,y) > y)
   x = abs(x);

x = abs(x) + max(x,y);
```

The formal parameters to a function are the names of the variables given in the parameter list for the function definition. Until the function is called, these parameters are just dummies, and do not represent anything. In **abs()**, the only formal parameter is **i**, while in **max()**, the formal parameters are **i** and **j**. When the function is called, a connection is made between the variables in the calling routine which are listed in the calling statement and the formal parameters. The actual variables in the calling routine are the actual parameters to the function. For the examples of function calls given above, **q** and **x** are actual parameters for the calls to **abs()**, while **x** and **y** are for **max()**.

There are a number of ways a programming language can make the connection between the formal parameters and the actual parameters. Two of the most popular are call by reference and call by value. In call by reference, the calling routine gives the function access to a pointer to the actual value of the variable, so any changes to the formal parameters in the function will be reflected in the actual parameters when control returns to the calling routine. Most FORTRAN implementations use this method of passing parameters. For call by value, the value of the variable in the calling routine is first copied to a new memory location before the function is called. Changes made to the formal parameters in the function will not be reflected in the actual parameters, since the function does not have access to the same memory location where the value of the actual parameter in the calling routine is stored. C uses this method for all nonarray (scalar) parameters.

Before a function call in C is made, any expressions in the actual parameter list are evaluated. Some examples of function calls involving expressions are:

j = abs(-8);
k = max((i+j),7);

This is true no matter what the type of the formal parameters. Syntactically, the elements of a function call list can be any legal C arithmetic or logical expressions.

When a function is called, the formal scalar parameters are replaced by the values of all the actual parameters in the calling statement. Copies of the actual parameters are passed to the function, rather than the actual parameters themselves. Assignments to the formal parameters within the function will not affect the actual parameters in the calling routine. This is a critical and important difference between C and other languages, like FORTRAN. In FORTRAN, all functions and subroutines are passed parameters by reference, so that a change in a scalar within a function will also change the value of the scalar within the calling function. This is decidedly not the case in C.

An example might help to clarify the situation. Consider the following function:

```
/*
*************************************
bump-bump up the value of x by 1 and print.
*************************************
*/

bump(x)

  int x;

{
    x = x + 1;
    printf("bump:%d.\n",x);
    return;
}
/*
end of bump
*/
```

Section 1.12-Functions in C   **33**

If we call **bump**() as in the following code fragment:

```
int y;

y = 0;
bump(y);
printf("y:%d.\n",y);
```

the result printed out is:

```
bump:1.
y:0.
```

Because the value of **y** is passed to **bump**(), incrementing **x** in **bump**() has no effect on **y** in the calling code.

If we want to have **bump**() increment **y**, we have to rewrite it so that it expects an integer pointer rather than an integer:

```
/*
*********************************
bump-bump up the value at the address
    pointed to by x and print.
*********************************
*/

bump(x)

  int *x;

{
    *x = *x + 1;
    printf("bump:%d.\n",*x);
    return;
}
/*
end of bump
*/
```

and call **bump()** with the address of **y** instead of **y** itself:

```
int y;

y = 0;
bump(&y);
printf("y:%d.\n",y);
```

Now the results are as desired:

bump:1.
y:1.

Arrays are an exception to the call by value rule. When a function is expecting an array as one of its formal parameters, what is actually passed to the function is a pointer to the first position in the array. This gives the function access to the beginning address of the array so address calculations for indexing can be performed properly. Changes made to array elements inside a function will show up in the caller, since the function has access to the array's beginning address in memory.

The following example illustrates these points:

```
/*
*****************************
sqvect-square elements in a double
       vector x with n elements.
*****************************
*/

sqvect(n,x)

  int n;
  double x[];

{
  for( n--; n >= 0; n--)
    x[n] = x[n] * x[n];

  return;
}
/*
end of sqvect
*/
```

sqvect() calculates the square of the elements in a double precision vector **x**. The parameter **n**

is the number of elements in the vector. The **for** loop first decrements **n**, so that it is the index of the last element in **x[]**, and loops until **n** is zero, which is the first index in **x[]**. If a call to **sqvect()** is made:

    sqvect(m,a);

then the elements of the variable **a** are replaced by their squares after the call, but the value of **m** remains unchanged.

Note that the parameter declaration for the array **x[]** does not require the size of the array to be specified. **x[]** could just as well have been declared:

    double *x;

since a pointer to the first **double** in the array is what the function will actually receive when it is called. Of course, somewhere in the calling code, **x[]** must be declared and dimensioned correctly.

Two-dimensional array parameters require more information in their formal parameter declarations than one-dimensional arrays. A two-dimensional array is stored in memory as a one-dimensional vector. When a function is expecting a two-dimensional array identifier as a parameter, the address of the first memory location in the one-dimensional storage form is passed, exactly as with a one-dimensional array. As discussed in Section 1.11, the address calculation to find an element with known row and column indices in a two-dimensional array requires that the number of columns be known. If the function doesn't know the column dimension, the address calculation cannot be made. Since the function has no way of finding out the column dimension from the calling function at compile time, the column dimension must be part of the formal parameter declaration in the function.

sqvect() can be rewritten to apply to a matrix:

```
/*
*****************************
sqmat-square elements of an n by m
      double matrix x.
*****************************
*/


sqmat(n,m,x)

  int n,m;
  double x[][MAXCOL];

{

  for( n--; n >= 0; n--)
    for( m--; m >= 0; m--)
      x[n][m] = x[n][m] * x[n][m];

  return;
}
/*
end of sqmat
*/
```

Here, **MAXCOL** is a constant defined using the C preprocessor, a topic that is discussed later in the chapter.

The **for** loops are arranged so that the dimension which varies the quickest in memory, namely the column dimension, is incremented the fastest. Looking back at how array address calculations are made, the base address for a row need only be calculated once per iteration of the outer loop with this arrangement. The inner loop must only perform a multiplication and add the result to the row base address. If the loops had been arranged the other way around, two additions and a multiplication would have been needed for each increment of the inner loop, since the base address of the row would need to be recalculated for each iteration. Smart compilers can often take advantage of this arrangement to speed up array indexing.

At first glance, the lack of an ability to change scalar parameters within a function might look inconvenient. However, it does avoid problems of unknowingly changing the value of an actual scalar parameter within a function, then having the effects mysteriously appear in some other part of the program, a bug often discovered after much searching. The use of the address operator in the calling function and a pointer within the function being called provides a way to consciously indicate that a scalar's value is changed by a function call.

1.12.1. Exercises

1.   Code a **main()** for the functions in the text. If you decide to do **sqmat()**, use a number for the column and row dimensions of the matrix. Do any initialization in a separate function, called **init()**.

2.   Write a function called **min()**, to find the minimum of two parameters. Test it with an appropriate **main()**.

3.   Extend **max()** and **min()** to take an integer vector as a parameter and return the maximum or minimum element in the vector. Now use a **float** or **double** vector.

4.   Try writing a function which takes an integer matrix as a parameter and writes out the elements of the matrix in neat rectangular form. Write a **main()** and **init()** which initialize a matrix and call the function, but declare the matrix in **main()** with a column dimension that is larger or smaller than the dimension in the output function. Do the elements of the matrix get printed out correctly?

1.13. Calculating the Factorial: An Example of Recursion

Functions in C can call themselves, either directly or through other functions. This ability is called recursion and is especially helpful in certain numerical problems which are naturally recursive. Unlike FORTRAN (but like Pascal) the C code for solutions to recursive numerical problems is therefore much cleaner and easier to implement than a FORTRAN solution would be, since the FORTRAN implementation must essentially simulate the recursive facility which C has.

A simple example of using recursion is the factorial function. The factorial of a nonnegative integer x is defined as:

$$f(x) = \begin{cases} 1 & \text{if } x = 1 \\ xf(x-1) & \text{if } x > 1 \end{cases}$$

The following C function **fact()** implements the factorial directly from the definition:

```
/*
***********************
fact-find the factorial of x.
***********************
*/

int fact(x)

   int x;

{
   if( x < 1 )
      return(ERR);

   else if( x == 1 )
      return(1);

   else
      return( x * fact(x - 1));
}
/*
end of fact
*/
```

ERR is an integer constant, distinguishable from a valid factorial, which is returned if the argument is invalid.

All recursive functions have some base case, like the exit condition of a loop, for which the recursion bottoms out and the function returns. A recursive function must have a path through it in which no direct or indirect call to itself is made. In **fact()**, the base case is **x == 1**. A recursive function without a base case is like an infinite loop.

As **fact()** also illustrates, the **return** statement at the end of a function may contain any legal C arithmetic or logical expression within the parentheses. The expression will be evaluated, including, as here, any function calls contained therein, and the result will be the return value of the function.

In terms of execution time, recursion is more expensive than a simple loop, because function calls take more time than an unconditional transfer to another location within the same function. The factorial function is probably simple enough that a loop could be substituted for the recursive call without affecting the clarity of the program significantly. However, some naturally recursive problems are much simpler to understand and program if recursive function calling is used. This is particularly true if the recursion is indirect, that is, one function calls another, which calls another, etc., until the first function is called again. Trying to simulate indirect recursion without actually doing a recursive function call usually involves a tangled mass of global variables which make the control flow difficult, if not impossible, to follow.

1.13.1. Exercises

1. Insert **printf()** statements at various points in **fact()** and try running it for a few different integers. What is the value of **x** on each call?

2. Rewrite **fact()** without using recursion. Which do you find easier to follow?

3. Remove the base case from **fact()** and try running it with an appropriate **main()**.

4. (Hard) Write a recursive function called **swap()** which takes a **double** array and two indices and swaps around the array between the indices.

1.14. Global Variables

Global variables can be defined in C. Unlike the local variables defined within a function, globals can be accessed anywhere. Global variables are declared by simply putting a variable declaration outside of a function definition. Typically, all the globals in a C program are defined immediately before the **main()**, isolating the variable definitions for global data in one spot.

For example, the following makes **charstring[]** available throughout a C program:

```
char charstring[MAXCHAR];

main()

{
    <main program>
}

<function definitions>
```

Again, **MAXCHAR** is a constant defined using the C preprocessor.

Any function defined in the same file after **charstring[]** can access **charstring[]** by merely referencing it. If, however, a function defines a local variable with the same name, the local definition will mask the global throughout the function and the global will be inaccessible. If **charstring[]** is referenced in a function before it is defined, or if it is used in a function whose definition occurs in a different file from where **charstring[]** is defined, then it must be declared before use and its variable declaration must be prefixed with the **extern** keyword.

For example:

```
main()

{
   extern char charstring[];

   <main program>

}

char charstring[MAXCHAR];

<function definitions>
```

Good programming practice suggests declaring a global using **extern** even if declaration would not legally be necessary. This gives a certain amount of self-documentation to the code by telling exactly where to look for the variable while you're debugging. It also avoids the problem of having a local variable definition inadvertently override that of a global. If globals are consistently declared **extern** before use in a function, then the declaration of a local with the same name as a global means that the global is not used in the function.

Note that, as in arrays passed to functions, the size of a global array need not be specified in the **extern** statement, since no space is being allocated by the compiler when an **extern** statement is encountered. For two dimensional arrays, the external declaration can be written as if it were the declaration for the formal parameter of a function, or the column dimension can be omitted entirely, since both the row and column dimensions are known globally from the original variable definition.

In general, global variables should be used only for data which is truly global to the entire program. Programs which use excessive numbers of global variables are difficult to debug and maintain, because the thread of data flow from function to function proceeds through a tangled interconnection of channels rather than through the well-defined interface of a function call. In Chapter 4, we'll discuss another type of global variable, the **static**, which allows the programmer to limit the number of functions that can reference certain data. This kind of global is more useful, because it allows information to be hidden from functions which shouldn't know about it.

1.15. Side Effects

A side effect is an event which occurs peripherally to the main effect of a computation. An example is a function which computes something, and, as a side effect, sets a global variable. A program which depends heavily on side effects can be extremely difficult to debug. Occasionally, however, side effects are desirable to smooth the flow of control.

The topic of side effects was briefly touched on in Sections 1.4 and 1.6, but the discussion was postponed until after functions and global variables were presented, so some concrete examples could be examined. In Section 1.4, side effects were mentioned in conjunction with the freedom of the compiler to rearrange mathematically associative expressions (like addition) and, in Section 1.6 they were presented with regard to short circuit evaluation of logical expressions.

As a simple example of how side effects could change the result of a computation, we could rewrite the **max()** function from Section 1.12 so that, in addition to returning the maximum of **i** and **j** as the value of the function, it also set the global variable **m** to the value of the minimum:

```
int m;

int max(i,j)

 int i,j;

{
  extern int m;

   if( i >= j)
   {
     m = j;
     return(i);
   }
   else
   {
     m = i;
     return(j);
   }
}
```

max() still returns the maximum, but as a side effect it determines the minimum too.

In the following code fragment, the value of **c** will differ depending on the order in which the addition subexpressions are evaluated:

```
extern int m;
int a,b,c;
int max();

m = 4;
a = 10;
b = 7;

c = m + b + max(a,b);
```

If the subexpression **m + b** is evaluated first, yielding 11, the function call to **max()** returns 10 and the resulting value of **c** is 21. If **b + max(a,b)** is evaluated first, the subexpression yields 17, since **max()** returns 10. As a side effect, however, **m** is set to 7 when **max(a,b)** is evaluated. When the entire expression is evaluated in this case, the result is 24. Since the compiler is free to rearrange the expression arbitrarily, the value of **c** after the statement is executed could be equally arbitrary. For this reason, side effects should be avoided in arithmetic expressions.

Side effects can also cause problems in logical expressions. If a logical expression is written so that evaluation could short out a subexpression having side effects, the code following the condition could be incorrectly executed. Consider the following code fragment, based again on the modified maximum function used in the previous example:

```
extern int m;
int a,b;

if( a > 0 || max(a,b) > 0)
    printf("The minimum is:%d.\n",m);
```

The logical subexpression after the **||** is not executed if **a > 0**, since the truth value of the entire expression can be determined without it. As a result, the value of **m** in the statement under control of the **if** may not necessarily be the minimum of **a** and **b**. This example can be rewritten so that **max()** always gets called by simply reversing the order of the logical subexpressions in the **if** condition:

```
if( max(a,b) > 0 || a > 0)
```

As a general rule of thumb, however, side effects involving global variables should be avoided wherever possible, since they are often difficult to trace and can easily lead to bugs. Other side effects, such as setting a variable as a result of executing a logical expression, can be useful if applied carefully. A common way of reading input in C programs is

```
while( (status = input(x,y,z)) != EOF &&
        status != ERR
     )
```

A side effect of analyzing the first logical subexpression in the **while** condition is to set **status**. In this case, the results of the side effect are clearly visible in the **while** condition, not obscured by a function call as in the case of the modified **max()** function in the previous examples. As long as the code within the **while** loop doesn't set **status** to **ERR**, expecting the call to **input()** to be skipped and the loop to be exited, this construct will work. The advantage of using a side effect is that the code for testing and setting **status** is neatly isolated within the **while** condition.

1.15.1. Exercises

1.   Program the two examples and see how they behave on your machine.

2.   Try some other arithmetic expressions with side effects, using parentheses to force the order of evaluation. Does your compiler reorder additions and multiplications despite the parentheses?

1.16. The C Preprocessor

In the definition of **sqmat**(), we used an identifier, **MAXCOL**, as the column dimensic for the two dimensional array. Similarly, the global character buffer **charstring**[] was dimensioned using the identifier **MAXCHAR**. Because the compiler needs to know how much space to allocate for the arrays and must generate the proper assembly language instructions for indexing, the constants **MAXCOL** and **MAXCHAR** must be defined at compile time. They cannot be global variables.

The C preprocessor provides a means of using symbolic constants such as **MAXCOL** and **MAXCHAR** and still having them defined at compile time. The C preprocessor is usually run before the C compiler and can be used to define constants and macro expressions which are substituted directly into the source code. A sharp sign (#) in column 1 of a line signals to the C preprocessor that it should interpret that line. Commands to the C preprocessor are not followed by a semicolon, since they are not C statements.

The array size constants can be defined with the following preprocessor statements:

```
#define MAXCOL     32
#define MAXCHAR    280
```

After these two lines, **MAXCOL** and **MAXCHAR** can be used anywhere throughout a file wherever the appropriate array bounds are needed. The advantage of the constants is that they have a certain meaning, indicating what role they play in the program, while the numbers are simply "magic numbers" which could just as well be replaced by some other number. Centralizing the array dimensioning information in one place is also an advantage when moving the program to a different machine. If more memory becomes available, the sizes of the arrays can be expanded without having to hunt through the entire program for places where the array was declared.

Macros can also be defined using the C preprocessor. A macro is a short piece of code that is substituted into source code by the C preprocessor before compilation, along with any arguments. If a particular function is very short, but is called often, then defining it as a macro with parameters keeps the code isolated but removes the function call each time that code must be executed. Since the code for a macro is inserted directly into the source, the overhead of a function call is saved. Another advantage of macros is that the types of parameters do not have to be specified, so a single macro can serve for **double**, **float**, or **int** arguments.

We can define a macro **DOUBLE**() which doubles its argument:

#define   DOUBLE(a)   ((a) * 2)

Parentheses were placed around the argument so that an expression can be substituted for **a** and still get the expected result. If parentheses had not been used, then the following code:

x = DOUBLE(i + j);

would not have evaluated correctly, since, after the macro is expanded and substituted into the source, the code would have looked like:

x = (i + j * 2);

Because multiplication is done before addition, the result would have been incorrect when the code was executed.

With the parentheses, the expanded code becomes:

x = ((i + j) * 2);

which evaluates correctly. Similarly, the parentheses were placed around the entire expression so that **DOUBLE**() can be used exactly like a function in an expression, and still have the code within the macro behave like a unit after substitution.

All constants and macros in this book are in upper case, to distinguish them from variables and function names. C identifiers are either all in lower case, or in lower case with a single upper case character at the beginning. This unambiguously differentiates between constants and macros defined using the preprocessor **#define** statement and C identifiers, so that someone reading the code doesn't have to puzzle over whether or not a particular identifier is a macro with arguments or a function.

The C preprocessor also provides a handy way of including the text from other files into source. The **#include** statement can be used to have a file inserted into the text at any point. The **#include** statement has the following form:

#include "globals.h"

The quote marks around the file name are required. Files included using the **#include** command usually have a **.h** (for header) extension.

When an **#include** is encountered by the preprocessor, the contents of the file with the indicated name are included directly in the source before compiling. **#include** files are a good way of keeping all global variables and defined constants together. Source files often have a line or two of **#include**s at the beginning, which merge in macros and constants defined in other files. **#include**s can be nested, so that files which are included can have **#include** statements in them for other files.

Some C library functions require that certain files containing system defined constants, macros, functions and variables be included in source which uses them. The C preprocessor knows the location of these files. If the file name is surrounded by angle brackets (<>) instead of double quotes, the preprocessor will consider the file name to be one of these system **#include** files:

#include <types.h>

A library function requiring that a system **#include** file be used should have the file's name listed in the documentation.

1.16.1. Exercises

1. Rewrite some of the functions in previous sections using preprocessor defined constants and macros.

2. Define a macro with one argument, called **TIMES**(), which multiplies its argument by a constant called **FACTOR** defined using **#define**. Write a program which prints the result of invoking the macro with an argument of type **int**. Try defining **FACTOR** before and after **TIMES**(). Does it make a difference? Try invoking **TIMES**() with the argument **FACTOR**. Does it work?

1.17. Summary

Since the primary purpose of this book is to introduce programmers to the idea of numerical software tools, the discussion of language features in this and future chapters is restricted to those aspects of the language which numerical programmers may find useful. C has other features which haven't been covered in this chapter. Some are used later in the book, and are discussed in more detail when they are needed. Others aren't that useful for numerical programming and will not be discussed. The references at the end of the chapter should be consulted by anyone who wants to know more about those aspects of C which aren't covered in this book.

An important part of using C which has just been mentioned in this chapter is the standard library of functions that comes with every C compiler. These functions do everything from input and output to the usual mathematical operations. Information on particular functions will be presented when they are first introduced. Further information on the standard function library is usually available with the C compiler documentation or in Sections 2 and 3 of *The UNIX Programmer's Manual*, Vol. 1.

1.18. References

There are a large number of introductory C books available on the market today. Most of them don't discuss numerical topics much, if at all, and are primarily oriented toward system programmers. Some which might be useful to the numerical programmer include:

1. *The C Programming Language*, by Brian W. Kernighan and Dennis M. Ritchie, Prentice-Hall, Englewood Cliffs, NJ, 240 pp., 1978.

    The "official" (though now somewhat dated) C handbook. Belongs on the desk of every serious C programmer. It contains a detailed language specification as an appendix, which should be consulted for very specific language questions.

2. *A Book on C*, by Al Kelly and Ira Pohl, Benjamin Cummings Publishing Co., Menlo Park, CA, 368 pp., 1984.

    A complete, up-to-date text on C, which covers everything, often with a more or less numerical view.

3. *A C Reference Manual*, by Samuel P. Harbison and Guy L. Steele, Jr., Prentice-Hall, Englewood Cliffs, NJ, 368 pp., 1984.

    A technical reference on C describing differences between compiler implementations and clarifying many of the language's fine points. Describes the post-1978 enhancements, like **void**.

A recent development in the evolution of the C language is a follow-on to C, called (appropriately enough) C++. C++ allows the programmer to define real types which have properties similar to the predefined types (like **int**, **float**, etc.). For example, programmer-defined types in C++ can have operations associated with them, exactly like the arithmetic operations associated with the **int** type in regular C. The programmer-defined data structures discussed in the next chapter do not have this property. Although C++ is not yet widely distributed, interested readers can learn about it in:

1. *The C++ Programming Language*, by Bjarne Stroustrup, Addison-Wesley, Reading, MA, 336 pp., 1986.

More information on the UNIX operating system can be obtained from *The UNIX Programmer's Manual*, Vol. 1 and 2. Vol. 1 has a number of sections, discussing various aspects of using the UNIX system. Section 1 contains information on commands, Section 2 is on system functions, and Section 3 contains the specifications for the C library functions. The content of the other sections varies, depending on the UNIX implementation. Vol. 2 contains in-depth information on certain commands and other topics from Vol. 1.

Beginning UNIX users may also want to check out:

1. *A User Guide to the UNIX System*, by Rebecca Thomas and Jean Yates, Osborne/McGraw-Hill, Berkeley, CA, 512 pp., 1982.

In addition, the August 1983 issue of *Byte* magazine was completely devoted to C, and the October 1983 issue was devoted to the UNIX system. Microcomputer users will want to check out these references, since they contain useful information on microcomputer implementations of C and the UNIX operating system.

Finally, the example illustrating the nonassociativity of floating point arithmetic was taken from:

1.   *Numerical Methods*, by Germund Dahlquist, Ake Bjoerck, and Ned Anderson, Prentice-Hall, Englewood Cliffs, NJ, 576 pp., 1974.

2

SOME VECTOR

AND

MATRIX TOOLS

The mathematical concepts of vectors and matrices are important in all numerical applications. A vector is an ordered set of numbers, often called a tuple, or n-tuple if the vector has n elements. Examples of two vectors which are different but contain the same elements are (1, 2) and (2, 1). Both are 2-tuples and they differ because the order of their elements is different. If a vector has n elements, then n is said to be the dimension of the vector. A vector can be arranged with all elements on the same line:

$$(1, 2, 3, 4, 5)$$

which is called a row vector, or with one element per line:

$$\begin{bmatrix} 1 \\ 2 \\ 3 \\ 4 \\ 5 \end{bmatrix}$$

as a column vector. In most cases, the difference is purely graphic, but some vector-matrix operations require that their data be either one or the other.

A matrix is a set of equal length vectors, arranged in a rectangular fashion:

$$\begin{bmatrix} 5 & 7 & 2 \\ 8 & 9 & 1 \end{bmatrix}$$

Depending on how you look at it, the matrix above consists of either two row vectors with dimension 3 or three column vectors with dimension 2. The dimension of the row vectors is called the row dimension of the matrix and the dimension of the column vectors the column dimension. If the matrix is square, i.e., the row dimension equals the column dimension, then we can just speak of the matrix dimension and do not have to specify which we mean.

Several special types of matrices are important in numerical applications. A square matrix is a matrix whose row and column dimensions are equal. The elements in a square matrix with equal indices are called the diagonal elements. A square matrix with ones on the diagonal and zeros elsewhere is called an identity matrix. Here is an example of a 3 x 3 identity matrix:

$$\begin{bmatrix} 1 & 0 & 0 \\ 0 & 1 & 0 \\ 0 & 0 & 1 \end{bmatrix}$$

A triangular matrix is a square matrix in which all the elements on one side of the diagonal are zero. A triangular matrix with all zero elements below the diagonal is called an upper triangular matrix, while one with all zero elements above is a lower triangular matrix. Here is an example of a 4 x 4 upper triangular matrix:

$$\begin{bmatrix} 3 & 4 & 5 & 6 \\ 0 & 1 & 8 & 7 \\ 0 & 0 & 2 & 4 \\ 0 & 0 & 0 & 1 \end{bmatrix}$$

From the programming point of view, vectors and matrices are stored in data structures called arrays. In Chapter 1, C **int**, **float**, and **double** array types of one and two dimensions were discussed. Data structures such as arrays are the first important ingredient in building numerical tool programs.

The second important ingredient is algorithms. Algorithms are recipes for manipulating data structures to obtain numerical results. In this chapter, three very simple algorithms for operating on vectors and matrices are discussed:

1. vector multiplication or the dot product,

2. vector and matrix addition,

3. solution of a system of linear algebraic equations, often called Gaussian elimination.

The three algorithms are implemented in three tool programs: **dotp**, **addm**, and **eqsolv**. As we develop these programs, we'll follow an implementation strategy called hierarchical or top down design. The hierarchical structure of most programs, with higher level functions calling lower level ones, suggests that the software design process should evolve along similar lines. Functions at the highest level are mapped out in very broad terms first and then refined into code, leaving the details of implementing the lower level functions until last. The tasks carried out by the lower level functions can be sketched out and the details filled in later. In this way, the design evolves from the general to the specific. Hierarchical design is particularly effective for simple programs, where the flow of data from one routine to the next follows a hierarchical pattern.

The rest of the chapter is concerned with developing the upper level command loops and the numerical code for the three programs. The command loops simply call lower level routines to fetch vectors and matrices from the standard input, **stdin**, call the numerical routines to operate upon the data, and call lower level routines to write the results to the standard output, **stdout**. We'll assume that we have a set of vector-matrix input and output (i/o) primitives available, which can read or write vectors and matrices to or from an arbitrary file. The implementation details of this interface are left for the next chapter, so that we can concentrate here on engineering the tool programs themselves.

The numerical routines implementing the actual algorithms for the dot product, matrix addition, and Gaussian elimination make up a very small fraction of the actual code. Indeed, this is usually the case for numerical software. Functions from a suitable numerical subroutine library could be substituted for those given here, and, in Chapter 4, interfacing non-C library functions to C programs is discussed. For this reason, we'll concentrate on the software engineering viewpoint in this chapter, stressing how numerical tools can be written which work well together, are easy to read and maintain, and can be reused over and over. Good modularization and a standardized vector-matrix i/o interface, which only needs to be written once, help us achieve these goals.

2.1. The Design of dotp

If $\vec{x}$ and $\vec{y}$ are two vectors of dimension n, then the dot product is simply:

$$p = \sum_{i=1}^{n} x_i y_i$$

If the dimensions of $\vec{x}$ and $\vec{y}$ differ, then their dot product is undefined. In our implementation of **dotp**, an attempt to multiply two vectors of differing lengths will be flagged as an error. The two vectors must be conformable for multiplication by having equal dimension.

Before writing any code, we want to thrash out a specification of exactly how **dotp** should work. Following the principle of hierarchical design, we'll sketch out the logic first and fill in the details later. In the initial design phase, a higher level, English-like pseudo-code is helpful.

Our design for **dotp** reads vectors from the standard input, calculates the dot product of vector pairs, and writes the result to the standard output. It continues to look for vectors until it encounters the end of file on standard input, in which case it terminates. In pseudo-code, the main routine for **dotp** looks something like this:

> **while**(not at end of file on standard in)
> {
>> Get equal length vectors from **stdin**;
>> Calculate their dot product;
>> Write the result to **stdout**;
> }

A lot has been left unsaid in this design. For example, what about errors? The pseudo-code written above assumes that the two vectors read from **stdin** have equal length so that they are compatible for calculating the dot product, and that they have been entered correctly by the user.

However, the pseudo-code does reveal the natural function-calling structure for the program. A function, **getvecs()**, is needed to fetch two vectors from **stdin**, making sure the vectors are compatible for calculating the dot product. If two incompatible vectors are input, **getvecs()** issues an error message and doesn't return until it has two vectors, or the end of file is read on the standard input. The details of exactly how the vectors are read from **stdin**, checking for correct syntax and reporting errors if the syntax is incorrect, will be left to a lower level function called **pgetv()**. Obviously, a function is also needed which takes as parameters two equal length vectors along with their mutual length, calculates the dot product, and returns the result. We'll call this function **vdotp()**.

In addition, **getvecs()** must return some kind of indication that the user wants to quit or that the end of the file from which vectors are being read has been reached. We'll use the constant **EOF** for this purpose. **EOF** is defined using a **#define** statement in the system **#include** file **stdio.h**. If the end of the file has been reached on **stdin**, **getvecs()** returns the constant **EOF** as the value of the function. If not, **getvecs()** returns the mutual length of the two vectors.

In general, passing a status variable back as the value of a function is a good way to indicate the function's termination status, and we'll use this technique in practically every function we write. Sometimes, as here, the return value can serve a dual purpose, either indicating that a particular status condition was reached or passing back some information, like the vector length, in addition to that returned through the other parameters. The calling routine can then take action, as required. A status variable should not be returned as a number (like -5) but rather as a preprocessor-defined constant which symbolically represents what caused the termination status. **EOF** is a good example. In Section 1.13, the constant **ERR** was used to indicate an error return from the function **fact()**. Other status variables which will appear at various points are **OK**, to indicate that the function returned successfully; **YES**, to indicate that the function performed some action requested by the calling routine; and **NO**, to indicate that a requested action was not performed.

Except for **EOF**, all these status constants are programmer defined and should be collected in one file, which we'll call **c.h**. This file can be included in every source code file by using the C preprocessor **#include** statement, so that all of our numerical routines will have access to the status constants. In addition, such things as macro definitions and the maximum sizes of arrays will be put in **c.h**, to avoid wiring "magic numbers" into programs.

Using **getvecs()** and **vdotp()**, the resulting main routine can be transformed into something resembling actual code:

```
while( (len = getvecs(v1,v2)) != EOF)
{
  x = vdotp(len,v1,v2);
  print x;
}
```

The return value from **getvecs()** is assigned to the variable **len** so it can be passed to **vdotp()**. At the same time, the assignment expression within the conditional part of the **while** lets us check if **EOF** was returned, and terminate the loop if so. Note that the assignment expression was surrounded by parentheses. If it had not been, **len** would have been assigned the result of the logical != operation, which is not what we want. As mentioned in Chapter 1, the assignment operator has the lowest priority of all operators, and is therefore always done last unless parentheses force a different evaluation order.

The main routine is now beginning to look more like C code. With the addition of the **main()** header, **#include** preprocessor lines, variable declarations, and the output statement, the code solidifies:

```
/*
****************************************
dotp-get two vectors from stdin, calculate dot
    product, write to stdout.
****************************************
*/


#include <stdio.h>
#include "c.h"

main()

{
  int status;
  int getvecs();
  double v1[MAXVEC],v2[MAXVEC];
  double vdotp();


    while( (status = getvecs(v1,v2)) != EOF)
```

```
        printf("%g\n",vdotp(status,v1,v2));

    exit(0);
}
/*
end of main
*/
```

The two preprocessor **#include** lines cause the files **stdio.h** and **c.h** to be included in the code for the main routine before it is compiled. In addition to the constant **EOF**, the file **stdio.h** contains definitions needed to use the standard i/o package, and it must be included whenever functions and variables from the standard i/o package, like **stdin** and **printf()**, are used. The standard i/o package functions are discussed in Sections 2.7 and 2.8. The constant **MAXVEC** gives the maximum vector length and is defined in **c.h**.

Notice that both **getvecs()** and **vdotp()** are declared as functions returning a particular type. In both cases, we're using the return values in later calculations. As mentioned in Chapter 1, we wouldn't have had to declare **getvecs()** because it returns an **int**. We did anyway, so a reader can look at the declarations and know how the return value of **getvecs()** will be used. For **vdotp()**, there is no need to assign the result to a variable before printing it, which is why the function call is made as an argument to **printf()**. C's call by value rule assures that the return value of the function call will be passed correctly to **printf()**. Because **vdotp()** returns a **double**, it can be used wherever a **double** can.

2.1.1. Exercises

1.   Try rewriting **main()** without the assignment statement in the **while** conditional or the call to **vdotp()** as an argument to **printf()**. Do you find the result any easier or less easy to read?

2.2. vdotp()

Implementing the dot product algorithm in C is also straightforward. Here is the code for **vdotp()**:

```
/*
****************************************
vdotp-return the dot product of double vectors
     x and y of dimension n.
****************************************
*/


double vdotp(n,x,y)

  int n;
  double x[], y[];

{
  double sum=0;

    while( --n >= 0)
     sum += x[n]*y[n];

    return(sum);
}
/*
end of vdotp
*/
```

Since **n** is passed by value to **vdotp**(), decrementing **n** in **vdotp**() has no effect on the value of **n** in **main**(); so no temporary is needed as an indexing variable. We've used the autodecrement operator in the prefix position to decrement **n** before comparing it with zero and using it as an index. **n** is the mutual dimension of the vectors. Because C has zero-based indexing, the last index in an array is always one less than the size of the array. Decrementing **n** makes it the correct index for the last element in the vectors. Similarly, the proper condition for terminating the loop is that **n** be less than zero, since the first element of the array has index zero.

vdotp() doesn't do much error checking, having assumed that the vectors being passed are valid **double**s, and that **n** is not larger than the size of the two arrays as declared in the calling function. Of course, this means that vector lengths must be checked before **vdotp**() is called. As we'll see in Chapter 3, the vector-matrix i/o interface function **pgetv**() catches oversized vectors.

A more serious problem could occur if any of the elements in **x**[] or **y**[] are large enough or small enough to cause an intermediate sum or product to overflow the machine's limits for double precision numbers. Checking for potential floating point overflow or underflow before a calculation is not as straightforward as checking for divide by zero errors, since the product of two numbers which are well within machine limits may cause a floating overflow. In addition, each element of **x**[] and **y**[] would need to be checked on each iteration of the **while** loop, slowing the dot product calculation considerably. Most minicomputer or mainframe implementations of C trap this type of error. Such support may not be available on all micro-computer implementations of C, however.

2.2.1. Exercises

1. The norm of a vector $\vec{x}$ is defined as the square root of the dot product of $\vec{x}$ with itself. Write a function **norm()** which takes the vector length and a **double** vector and returns the norm as the value of the function. To calculate the square root, use the C library **double** function **sqrt()**. **sqrt()** takes a single positive **double** argument and returns the square root of the argument as the **double** value of the function.

2.3. Initialization and the Extended Assignment Operator

vdotp() also illustrates two other features of C that can make program code more compact by reducing the number of extraneous assignment statements. These features are initialization within a declaration statement and the extended assignment operator. In C, any scalar variable can be initialized by simply following the variable's name in the declaration with an = and the value to which the scalar should be initialized. The running sum variable **sum**, used in the dot product calculation, was initialized in this way.

Both local and global variables can be initialized in declaration statements. For global variables, the initialization is done only once, before the program starts executing. For locals, however, the initialization takes place every time the function is entered, since a new copy of a local variable is made each time the function is called. In Chapter 4, we'll discuss initialization more thoroughly, particularly how to initialize more complex data structures.

The extended assignment operator += removes the need to put an identifier on both sides of an assignment, if the effect of the statement is to increment the identifier using some arithmetic operation. The statement under control of the **while** loop in **vdotp()** is an example. It's equivalent to the statement:

sum = sum + x[n] * y[n];

An extended assignment statement is formed by putting any of the arithmetic operators directly to the left of the assignment operator =. The right-hand side of the assignment expression is first evaluated, then the result is used in an arithmetic expression with the arithmetic operator and the variable on the left side. The result of the arithmetic operation is stored back in the variable.

The extended assignment operator can be used inside an expression exactly like the assignment operator itself, but, as with the assignment operator, parentheses may be used to force precedence. Although it can save some typing, too intensive use of the extended assignment operator can make code cryptic and hard to read. For this reason, we'll confine its use to cases where a running sum or product is being calculated and the meaning is obvious.

2.4. getvecs()

Because the dot product algorithm concisely outlined what needed to be done, pseudo-code for **vdotp()** wasn't necessary. That is not true of **getvecs()**. As mentioned previously, **getvecs()** must fetch two vectors from **stdin**, using the lower level routine **pgetv()**, and check if the two vectors are of equal dimension. If not, **getvecs()** should issue an error message and look further, until either **EOF** is returned from **pgetv()** or two vectors compatible for vector multiplication are obtained.

With this in mind, here is the pseudo-code for **getvecs()**:

```
Set the length of vector1 and vector2 to 0;

while(we don't have two equal length vectors)
{
    Get vector1 from stdin;

    if( at end of file)
    {
        Terminate the loop
    }
    else
    {
        Get vector2 from stdin;

        if( at end of file)
        {
            Terminate the loop
        }
        else if( length vector1 != length vector2)
        {
            Write a message to stderr;
        }
    }
}

if( at end of file)
    return(EOF);

else
    return( length of vector1);
```

The actual code follows directly from the pseudo-code with one minor and one major modification:

```
/*
**************************************
getvecs-get two equal length double vectors
 from stdin, return dimension or EOF.
**************************************
*/

int getvecs(v1,v2)

  double v1[],v2[];

{
  int len1=0,len2=0;
  int pgetv();

    do
    {
     if( (len1 = pgetv(stdin,stdout,v1)) == EOF)
       break;

     else if( (len2 = pgetv(stdin,stdout,v2)) == EOF)
     {
       fprintf(stderr,
            "dotp:unmatched vector at end of file.\n"
          );
       break;
     }
     else if( len1 != len2 )
       fprintf(stderr,"dotp:vector lengths unequal.\n");

    }while( len1 != len2);

    if( len1 == EOF || len2 == EOF)
      return(EOF);

    else
      return(len1);
}
/*
end of getvecs
*/
```

The major modification is the use of two new control statements, the **do-while** loop and the **break** statement. The **do-while** loop is similar to the **while** loop, except that the looping condition is checked at the bottom of the loop instead of at the top. The **break** statement provides a way of breaking out of a loop without using a **goto**. Both statements are discussed in the next two sections.

As a minor modification, an extra error test was put into the loop body, to catch unmatched vectors at the end of file. This message is more informative to the user than the vector length message which would otherwise have been printed. The C standard i/o package function **fprintf()** is used to print the error messages to the standard error output, **stderr**. **fprintf()** works exactly like **printf()**, except a **FILE** pointer parameter must be given before the format string. The standard error output, like **stdin** and **stdout**, is opened when the program begins executing and, if no other arrangements are made, messages written to **stderr** appear on the terminal screen. **FILE** pointers and i/o using the standard i/o package are discussed in Section 2.8.

When **pgetv()** returns, the **double** array passed as the last parameter is guaranteed to contain a double precision vector with no more than **MAXVEC** elements, unless the end of file was reached. The number of elements (or **EOF** if the end of file was reached) is returned as the value of the function. **pgetv()** is also passed **FILE** pointers, instead of just reading directly from **stdin**, because in later chapters it will become necessary to read vectors from and write vectors to files which are not associated with the standard i/o channels.

Since **pgetv()** has the task of returning vectors from an input file, why have we included an output **FILE** pointer in the parameter list? Any errors could certainly be written to **stderr**, as **getvecs()** does. The reason will become clearer in the next chapter, where the data interface for exchanging vectors and matrices between tool programs is discussed, but, for now, the following argument may help to justify this seemingly contradictory design decision. If **dotp** is used together with other programs, like **addm**, it may be necessary to pass vectors on through **dotp**. **dotp** shouldn't touch these vectors but rather simply write them out as soon as they are read in, with no further processing. If we want to isolate all the code for handling the data exchange interface in **pgetv()** (and its matrix counterpart **pgetm()**), **pgetv()** must know where to write these vectors. The extra **FILE *** parameter serves this purpose.

Error handling for vector input in **dotp** and other tools is designed so that lower level routines catch and report errors and only return when they have correct data. The error messages tell the user exactly what caused the error, so it can be corrected on the next input attempt. Such a design gives the user feedback when errors are made, and allows the user to recover from errors gracefully. This would not be the case if an error message simply caused the program to terminate. The user would then be required to restart the program and redo any work to set up the calculation.

An alternative design is to have the routine finding the error return an error code and leave the reporting up to the calling function. This design is more appropriate for functions which do lower level processing, such as numerical algorithms. The calling function can then gather the error codes or abort processing and report the error. In Section 3.4, the topic of error handling is discussed in more detail.

2.5. The do-while Loop

The **do-while** loop is similar to the **while** loop, except that the condition for looping is not checked before the loop is entered. The **do-while** loop has the following syntax:

```
do
  statement;
while(expression);
```

The statement can be a compound statement and is executed until the value of the expression becomes zero. As in the **while** loop, the expression must be changed within the body of the loop.

A **do-while** loop was used instead of a **while** in **getvecs()** because the condition to continue looping, namely that two equal length vectors have not yet been fetched, will always be true at the top of the loop. The **while** loop would require initialization conditions which seem somehow unnatural:

```
len1=0;
len2=1;

while( len1 != len2 )
{
  body of loop
}
```

An alternative would be to introduce an additional status variable upon which the **while** test could be made. The status variable could be set to **NO** before the loop is entered, then changed to **YES** when two suitable vectors are fetched. But why introduce an additional variable? The **do-while** code emphasizes what the looping condition should be, namely continue looping until two equal length vectors have been fetched.

The **do-while** loop is less useful in general than the **while** or **for**; however, it can make code less complicated when the condition for looping is guaranteed to be true upon entry to the loop and changes only during the course of executing the loop body. Loops which must execute at least once regardless of conditions when the loop is entered, as in the present case, are an example. If a loop has some entry condition which must be fulfilled before the loop is entered, then a **for** or a **while** loop is more appropriate.

2.6. break and continue

The **break** statement allows a program to gracefully break out of loops without using **goto**s and labels. When the **break** is encountered, control moves to the first statement after the immediately surrounding loop. If more than one loop surrounds a **break**, then additional **break** statements are required to break out of those loops.

The code for **getvecs**() illustrates very nicely how useful the **break** statement is. Without the **break** statement, the looping condition in the **do** loop becomes considerably more complicated:

>**}while(len1 != EOF && len2 != EOF && len1 != len2);**

Not only that, but the relational expressions in the looping condition must be written exactly in the order shown. If the **len1 != len2** condition is evaluated before **len2 != EOF**, the loop will continue when **pgetv**() returns **EOF** on the second call but not the first. Since, by short circuit evaluation, **&&** fails in C as soon as one term is zero, putting the terms in the order given will ensure that the loop terminates correctly. Even better, the **break** statement avoids having to think about a potentially bug-prone conditional, and makes the code easier to read too.

While we're discussing transfer of control within loops, we might as well mention **break**'s cousin, the **continue** statement:

>**while**(expression)
>{
>   . . .
>   **continue**;
>   . . .
>}

The **continue** statement causes control to transfer to the point in the loop immediately before the loop continuation condition is checked, in the **while** and **do-while** loops, or before the incrementing statement is executed, in the **for** loop. The net effect is to skip the entire portion of the loop after the **continue**, transferring to where the establishment of the conditions for the next iteration takes place.

The **continue** statement is seldom used, but it can sometimes be very useful. More complicated loops may require that parts of the loop be skipped after one or two iterations, because the values of certain variables have changed. The only alternative to **continue** for these cases is to use a **goto** and a label.

2.7. printf()

In Chapter 1, several short sample programs used the function **printf**() to write out an integer. The **printf**() function in **dotp** is used to write out a double precision real. In general, **printf**() can be used to write numerical, character, or string data to the standard output. In this regard, it's a little like the PRINT statement in FORTRAN. **printf**() is called in the following manner:

>**printf**(<format>, <param1>, <param2>, ..., <paramn>);

printf() formats <param1> through <paramn> according to the specification in the string <format> passed as the first parameter and writes the result to **stdout**. The number of arguments is arbitrary, but the number and types of the arguments must match the format specification or the result will be unpredictable.

The format string contains two types of information, text and conversion specifications. Text is simply copied verbatim to the standard output. Examples include the messages printed out in Chapter 1. Conversion specifications cause the parameters to be formatted and written to the standard output at the place in the output text where the conversion specification appears. The **%g** string in **dotp** is an example. A conversion specification consists of the percent character (**%**), followed by some optional formatting information, followed by the conversion character itself. The conversion character must be one of the following:

d   The argument is an integer and should be printed as a decimal (base 10) integer.

o   The argument is an integer and should be printed as an octal (base 8) integer.

c   The argument is a character or integer in the range 0-177 octal (0-127 decimal). It should be printed as an ASCII coded character.

s   The argument is a character string. Characters from the string are printed until a null character is reached or the number of characters specified in the optional precision information between the **%** and the **s** is achieved.

e   The argument is a **float** or **double** and is converted to exponential notation: [-] x.yyyyyyE+-zz, with the number of y's being specified by the precision. The default precision is 6.

f   The argument is a **float** or **double** and is converted to decimal notation: [-] xxx.yyyyyy, where the number of y's is specified by the precision. The default precision is 6. The precision does not determine the number of significant digits.

g   **%e** or **%f** format is used, depending on which one is shorter. Nonsignificant zeros are suppressed.

The optional formatting information which may occur between the **%** and the conversion character can include the following, in order of appearance left to right after the **%**:

1.   A minus sign, specifying that the converted argument should be written in the leftmost side of the output field (left adjusted). If no minus sign appears, the default is to right adjust the output.

2.   A digit string, which specifies a minimum field width. The number, character, or string converted from the argument will be printed in a field at least that wide, and wider, if necessary. There is no ***** output if the number doesn't fit in the field, as in FORTRAN. The converted argument is padded on the left or right, depending on whether left adjustment was requested, if it is smaller than the field width. The padding character is normally a blank, unless the digit string specifying the field width begins with a 0, in which case zeros are used.

3. Another digit string, separated from the field width by a period (.). This number specifies the maximum number of characters to be printed from a string, or the number of digits printed to the right of a decimal point if the argument to be converted is a **float** or **double**. The number is ignored if the argument is an **int** or **char**.

Finally, the % character itself can be printed by preceding it with another %, as %%.

Additional examples of **printf()** will appear throughout the book; however, here are a few mixing text and numbers. The array **string[]** contains the character string "**The completion rate was**":

```
char string[MAXBUF];
int i;
double f;

printf("The completion rate was %-12.5E%%.\n",f);
```

Here the minus sign indicates left adjustment in a field 12 digits wide, with 5 digits after the decimal point. The result is:

The completion rate was 1.00147E+1 %.

In the next example, the **%s** specification causes the string variable to be inserted into the output text, instead of putting the string into the format itself. The lack of a minus sign in the optional formatting information for the variable **f** indicates right adjustment. The use of a lower case **e** in the format string causes a lower case **e** to be output in the exponential notation:

```
printf("%s %12.5e%%.\n",string,f);
```

The resulting output is:

The completion rate was   1.00147e+1%.

Finally, an array value can be designated in the output by having the integer value of **i** substituted into the brackets. Default field widths and precisions are used:

```
printf("s[%d]:%f\n",i,s[i]);
```

If **s[i]** is 1.25, the output would be:

s[2]:1.250000

Notice that each format string contains a newline character at the end. Without a new-line, the next output appears immediately after the previous string on the same line.

2.7.1. Exercises

1. Write a program which tests **printf()**'s formatting capabilities. Use a number of different options, including left and right justification, specifying how many charac-ters to output from a string, no newline at the end of a format string, etc.

2.8. File I/O Using FILE Pointers

The **printf()** function discussed in the previous section allows writing to **stdout** only. File i/o with **stdin**, **stdout**, and **stderr** as well as with other files can be accomplished using the standard i/o package found in the C library. **FILE** pointers, like **stdin**, **stdout**, and **stderr**, are used similarly to FORTRAN logical unit numbers for performing i/o with peripheral devices. **stdin**, **stdout**, and **stderr** have type **FILE ***. The **FILE** keyword is defined using the C preprocessor **#define** statement in **stdio.h**. However, it is not necessary to know the detailed definition of **FILE** in order to use **FILE** pointers.

FILE pointers are passed as parameters to the standard C library functions **fprintf()**, **fgets()**, **getc()**, and **putc()**. Each of these functions has a counterpart which only does i/o with **stdin** or **stdout**. These counterparts are **printf()**, **gets()**, **getchar()**, **putchar()**.

stdin, **stdout**, and **stderr** are already opened when the program begins executing. Before another file can be accessed using a **FILE** pointer, however, it must be opened using the C library function **fopen()**:

```
            FILE *fd, *fopen();
            char *filename, *mode;

            fd = fopen(filename, mode);
```

fopen() makes a connection between the external file and the program. **filename** is a character string containing the name of the file as it is known to the operating system. The possible character strings that can be substituted for **mode** differ, depending on the flavor of the UNIX system you're running, but three which come with every version of C are:

"**r**" file is to be opened for reading only,

"**w**" file is to be opened for writing only,

"**a**" file is to be opened for appending.

For example:

```
            fd = fopen("myfile","r");
            file = fopen(buf,"w");
```

If an attempt is made to open a file read only and the file is not there, **fopen**() returns **NULL**, a null pointer value defined in **stdio.h**. A file opened for write or append access will be created, if possible. An existing file opened with "**w**" for **mode** is truncated and the old contents of the file are thrown away. After read or write access has been established using "**r**" or "**w**" for **mode**, information can be read or written starting at the beginning of the file. Append access causes any old information in the file to be maintained, and new information to be written at the end of the file after the old. If, for some reason, a file cannot be created, or some other error occurs, **fopen**() returns **NULL**.

The function **fclose**() breaks the connection between the file and the program:

> **fclose(fd)**

fclose() should be called when no further i/o operations are necessary with the file, since there is a limit on the number of files a program can have open simultaneously. When a program exits, however, all files are automatically closed.

Once a file has been opened, **fprintf**() and its comrades can be called to put information into the file or to fetch information from it:

> **FILE *fd;**
> **char buf[MAXBUF], c;**
> **char *fgets();**
> **int maxlen=MAXBUF;**
> **double x;**
>
> **fprintf(fd,"%c %d %g\n",c,maxlen,x);**
> **fgets(buf,maxlen,fd);**
> **c = getc(fd);**
> **putc(c,fd);**

Here are the same examples using the standard i/o counterparts:

> **printf("%c %d %g\n",c,maxlen,x);**
> **gets(buf,maxlen);**
> **c = getchar();**
> **putchar(c);**

fprintf() works exactly like **printf**() except that the results show up in the file associated with the **FILE** pointer **fd** rather than at **stdout**. The format string follows the same rules as **printf**().

fgets() reads characters from the current position in the file associated with **fd** until either **maxlen-1** characters have been read, a newline character is encountered, or end of file is reached, whichever comes first. The result is returned in **buf** as a null terminated string, with the newline, if any, as the last character. On end of file, **fgets**() returns **NULL**. **gets**() works exactly like **fgets**(), except input is taken from **stdin**, so no **FILE** pointer argument is required.

In addition, just to make things confusing, **gets()** drops the newline character before returning the null terminated string. If end of file was not encountered, both **gets()** and **fgets()** return a pointer to **buf** as the value of the function. A typical way of using **fgets()** or **gets()** in a C program is within a **while** loop:

```
FILE *fd;
char buf[MAXBUF];
char *fgets();

    while( fgets(buf,MAXBUF,fd) != NULL )
    {
      do something with buf
    }
```

This kind of construct occurs frequently in routines which do input, as we'll see in the next chapter.

getc() gets a single character from the file associated with the **FILE** pointer **fd**, and returns that character as the value of the function. If the end of file is reached, **getc()** returns **EOF**. Similarly, **putc()** puts the character **c** to file **fd**. Their counterparts for the standard i/o channels, **getchar()** and **putchar()**, operate exactly the same, except no **FILE** pointer argument is required and i/o is done with **stdin** and **stdout**. Neither **getc()** nor **getchar()** was declared as a functions because neither is a function. Both are macros, defined using the preprocessor **#define** statement in **stdio.h**, as are **putc()** and **putchar()**. If they had been declared, the compiler would signal an error, since the preprocessor would substitute the code directly in line.

2.8.1. Exercises

1. Write a program called **stash**, which reads lines from **stdin** and squirrels them away in a file called **stash.out**.

2. Write the companion program to **stash**, called **unstash**, which reads lines from **stash.out** and writes them to **stdout**.

3. Write a program called **vowel**, which counts the number of vowels occurring in a sequence of characters read from **stdin**, and writes a table out afterwards with each vowel and the number of times it occurred.

2.9. The Matrix Addition Algorithm

The next numerical tool which we'll develop is a program called **addm**. **addm** does matrix addition. If A is an m x n matrix and B is an m x n matrix, then the matrix sum is:

$$C = A + B = \begin{bmatrix} a_{11}+b_{11} & & a_{n1}+b_{n1} \\ \cdot & & \cdot \\ \cdot & \cdots & \cdot \\ \cdot & & \cdot \\ a_{m1}+b_{m1} & & a_{mn}+b_{mn} \end{bmatrix}$$

The sum matrix is constructed by adding elements at the same position in the two summand matrices.

Like the dot product, the two matrices must be conformable for addition, with the same number of rows and the same number of columns, though the matrices needn't be square. Note that matrix addition is commutative like the addition of scalars. Thus, if A and B are both m x n matrices, $(A+B)$ equals $(B+A)$.

2.10. Programmer-Defined Data Structures and the matrix Type

In general outline, the hierarchical structure of **addm** is virtually identical to **dotp**. A main routine loops, calling a routine to get two matrices conformable for addition from **stdin**, a routine to add them, and a routine to write the matrices to **stdout**. In fact, by substituting arrays with two dimensions for those with one and using a routine (**mput()**) to write out matrices instead of **printf()**, the code for the main routine could be simply carried over from **dotp**:

```
/*
************************************************
addm-fetch two matrices from standard input, calculate
        their matrix sum, and write the result to the
        standard output.
************************************************
*/

#include <stdio.h>
#include "c.h"

main()

{
  int row,col;
  int getmats();
  double mat1[MAXVEC][MAXVEC],
        mat2[MAXVEC][MAXVEC],
        mat3[MAXVEC][MAXVEC];
```

```
         while( getmats(&row,&col,mat1,mat2) != EOF)
         {
           madd(row,col,mat1,mat2,mat3);
           mput(stdout,row,col,mat3);
         }

         exit(0);
}
/*
end of main
*/
```

There are a couple of important differences, however. The algorithm for matrix addition requires that the number of rows and the number of columns match in the two matrices. This is similar to the requirement that the dimensions of the two vectors for the dot product be equal. **getmats()** must check that this is true. Because two integers must be returned from **getmats()** instead of one, integer pointers must be passed as arguments, since, otherwise, the values returned will not be available in **main()**. The row and column dimensions must be passed to **madd()**, which does matrix addition. The dimensions of the resulting matrix must also be passed to **mput()**, which will write the matrix to **stdout**. Unlike FORTRAN, C has no equivalent of the variable dimension array in a function, so **madd()** would have to be written with the column dimensions of the three arguments specified.

Although the above implementation is adequate, and would be a typical one for FOR-TRAN, it is messy in several respects. Matrices are not really just two-dimensional arrays, but rather, two-dimensional arrays with a certain row and column size. Because FORTRAN lacks any data structures more complicated than arrays, this information is usually passed around through variables in addition to the matrix, resulting in subroutine calling statements which are often long and not very easy to read. Such code is also prone to bugs. Long subroutine calls make it easy to overlook a typo that may result in the wrong row or column dimension being passed as a parameter. Of course, the row and column dimensions could be put into a one-dimensional array which could be passed around along with the matrix, but the fundamental problem of how to "bind" the row and column dimensions to the matrix would not have been solved.

C has the perfect solution: a programmer-defined data type called a **struct**. A **struct** is essentially an aggregate object, containing several members having some kind of common, functional relationship. The declaration:

```
         struct matrix
         {
           int row,col;
           double m[MAXVEC][MAXVEC];
         };
```

defines a **struct** template. The **matrix struct** can be used for declaring variables to hold matrices and their dimensions in one object. The **double** array **m[][]** is used to hold the matrix elements while the **int**s **row** and **col** hold the dimensions.

In general, the syntax of a **struct** template declaration is:

```
struct <identifier>
{
  <member declarations>
};
```

The member declarations are regular C variable declarations for either predefined or programmer-defined (i.e., **struct**) data types. A **struct** can even contain a member with a variable that has the same type as itself.

Notice that this definition doesn't allocate any memory for a **struct**, but merely defines a template and a name, which can serve to declare memory allocation for a **struct** later in the program. To actually allocate memory, the variable name can either be inserted between the closing curly brace and the semicolon in the template definition, or the variable can be declared exactly like the declarations for the predefined types:

struct matrix mat1,mat2,mat3;

This declaration can be substituted for the declaration of the three double arrays at the beginning of **main()**.

structs can be used in variable declarations exactly like the predefined data types. Arrays of **struct**s, pointers to **struct**s, or arrays of pointers to **struct**s can be declared. Pointers to **struct**s are particularly useful when passing **struct**s between functions as parameters. Some earlier versions of C wouldn't allow a **struct** to be passed as a parameter to a function, or a function to return a **struct** as its return value. This restriction can be circumvented by simply passing around pointers to **struct**s. Although the restriction has now been removed, passing a **struct** to a function tends to be considerably more expensive than passing a pointer. If a **matrix struct** were passed to a function, for example, every one of the elements in the matrix, and the row and column variables would have to be pushed onto the function call stack, a total of **MAXVEC** x **MAXVEC** + 2 operations. If only a pointer is passed, then only a single address need be pushed. For this reason, and to maintain backward compatibility with older C compilers, only pointers to **struct**s will be passed as actual parameters to functions in this book.

The members in a **struct** are accessed by appending a period (.) to the **struct** variable name, followed by the name of the variable member. Given the declaration of the three matrices above, the following examples illustrate read and write access to member elements:

```
mat.m[1][5] = 20.0;
i = mat.col;
mat.row = mat.col;
```

To access a member using a pointer to a **struct**, the dereferencing operator * and the period operator . could be used:

```
struct matrix *mptr;
double x;

    mptr = &mat;
    x = (*mptr).m[5][5];
```

However, there is an alternative way to do this, using the arrow operator (->). The arrow operator is positioned similarly to the period, between the **struct** pointer variable name and the member name, for example:

```
    x = mptr->m[5][]5];
```

which is equivalent to the above. The arrow operator is preferable to the period because no parentheses are required. Note that if we omit the parentheses when using the period, we get:

```
    x = *mptr.m[5][5];
```

which is completely incorrect. It is an attempt to dereference **mptr.m[5][5]**, which isn't a pointer to a **struct**. In fact, **mptr.m[5][5]** isn't even the correct use of the dot operator, since **mptr** isn't a **struct** but rather a pointer to a **struct**.

In earlier (pre-1978) versions of C, **struct**s could not be assigned as units; however, most later versions now allow block assignment. For example, if **imat** contains an identity matrix with the **row** and **col** members also initialized, the statement:

```
    *mptr = imat;
```

would initialize the **struct** pointed to by **mptr**. Other operations, such as logical comparisons, are not permitted, however. A **struct** assignment has exactly the same effect as assigning any other variable, namely the contents of **imat** are copied into ***mptr**.

structs improve program readability, but the C **typedef** facility is even more useful for indicating the function of data within a program. **typedef** can be used to declare a new type keyword. The new keyword becomes a type name and can be used like the built-in keywords **int**, **double**, etc. A **typedef** declaration has syntax:

```
    typedef <defined type> <identifier>
```

where <defined type> is one of the C predefined types (**int**, **char**, **float**, **double**), including pointers and arrays, a programmer-defined type (**struct**), or a type previously defined using **typedef**. The identifier becomes the new type identifier. For example, the following **typedef** statement allows the new keyword **complex** to be used for identifying certain two-dimensional arrays as holding complex numbers:

typedef double complex[2];

complex z;

The second statement will be treated exactly as if **z** had been declared as a two element **double** array. The advantage is that the purpose for which **z** is used in the program is clearer than if **z** had simply been declared as a **double** array.

Applying this to the matrix declaration, a **matrix** type can be defined as follows:

```
typedef struct
{
  int row,col;
  double m[MAXVEC][MAXVEC];

} matrix;
```

The declaration:

matrix m1,m2,m3;

can now be used instead of using the **struct** declaration itself.

Both **struct** template and **typedef** declarations are only valid in the files where they appear. Local and global variables and function parameters of a **struct** or **typedef** type can be declared in a file only if the **struct** template or **typedef** definition appears in the file. A common way of sharing **struct** templates and **typedef** definitions throughout a program whose source code occupies several files is to put the **struct** templates and **typedef** statements into a header file, like **c.h**, then include this header file in all source modules which need to use the definitions, using the preprocessor **#include** statement.

C is not as strongly typed a language as Pascal, and the **typedef** statement is more an optional feature, to be used where necessary to improve readability, rather than an absolute requirement. Some Pascal programmers have a tendency to overuse typing, making everything in sight a new type. In general, excessive use of typing tends to defeat the purpose, requiring someone who is reading the code to check back to the function header or global declarations for the type of every new variable. In addition, proliferating types excessively causes the amount of code to increase. An algorithm that could be applied to some generalized data structure will require a function for each specific type, since the arguments to the function must be typed. A good rule of thumb for defining a new type is that data which belongs to some logically cohesive group (like matrices) is a candidate for a new type. If used with discretion, **typedef** can lead to more readable code.

With the **matrix** type, the **main**() function can be rewritten:

```
/*
*************************************************
addm-fetch two matrices from standard input, calculate
       their matrix sum, and write the result to the
       standard output.
*************************************************
*/

#include <stdio.h>
#include "c.h"

main()

{
  int getmats();
  matrix mat1,mat2,mat3;


    while( getmats(&mat1,&mat2) != EOF)
    {

       madd(&mat1,&mat2,&mat3);
       fputm(stdout,&mat3);
    }

    exit(0);
}
/*
end of main
*/
```

Pointers to **mat1** and **mat2** must be passed to **getmats**(), since **getmats**() must be able to access the addresses of the matrices, so that the input values can be deposited there. Similarly, **madd**() must get a pointer to **mat3** because, on return, **main**() needs to pass the sum matrix to **mput**(). Remember, passing a **struct** to a function is exactly like passing an integer: the value of the **struct** is copied and only that copy is available to the function, not the **struct** itself. Modifications to a **struct** can only occur if a pointer to the **struct** is passed, so the function can get at the **struct** through its address.

2.10.1. Exercises

1.   Define a type called **vector** using **typedef**, which bundles together a vector and its length. Extend the type so that the vector records whether it is a column or row vector.

2.   Rewrite **vdotp()** using the vector type. Do you think the result merits further use? Justify your answer.

2.11. getmats()

In outline, the control flow of **getmats()** is exactly the same as **getvecs()**. The code follows by replacing **pgetm()** for **pgetv()** and substituting the appropriate test for matrix addition conformability:

```
/*
**************************************
getmats-get two correct matrices from stdin,
        checking for addition conformability.
**************************************
*/

int getmats(mat1,mat2)

  matrix *mat1,*mat2;

{
  int status = NO;
  int pgetm();

    do
    {
     if( (status = pgetm(stdin,stdout,mat1)) == EOF)
       break;

     if( (status = pgetm(stdin,stdout,mat2)) == EOF)
     {
       fprintf(stderr,
           "addm:unmatched matrix at end of file.\n"
           );
       break;
     }
     else if( mat1->col != mat2->col ||
```

```
                            mat1->row !=mat2->row
                        )
              {
                fprintf(stderr,
                        "addm:column or row sizes don't match.\n"
                        );
                status = ERR;
              }
              else
                status = YES;

        }while( status == ERR);

        return(status);
}
/*
end of getmats
*/
```

Using the status variable isolates the check for equal row and column dimensions to one branch of the multiple **if-else** statement, rather than requiring an extra check in the looping condition. The loop terminates when two matrices with equal row and equal column dimensions are fetched. The return value of **getmats()** is either **EOF** or **YES**, indicating either that the end of file was reached or that two matrices were fetched. Unlike **getvecs()**, the row or column dimension is not returned as the value of the function, but within the returned **matrix struct**.

2.12. madd()

The algorithm for matrix addition discussed in Section 2.9 can be translated into pseudo-code:

```
for( each row )
    for( each column )
    {
        Add the two matrix elements and record the result in the third;
    }

    Set the number of columns and rows in the third matrix;

    return(OK);
```

Using the **matrix** type, the pseudo-code becomes the following C function:

```
/*
***************************
madd-add matrices m1 and m2.
    Put the result in m3.
***************************
*/

int madd(m1,m2,m3)

  matrix *m1,*m2,*m3;

{
  int i,j;

    for( i=0; i < m1->row; i++)

      for( j=0; j < m1->col; j++)
        m3->m[i][j] = m1->m[i][j] + m2->m[i][j];

      m3->row = m1->row;
      m3->col = m1->col;

      return(OK);
}
/*
end of madd
*/
```

Again, little error checking is done, since we've assumed that mismatched matrices have been filtered out in **getmats()**. As in Chapter 1, the order of the **for** loops was arranged to take advantage of the order in which C stores arrays. Note that the arrow operator, not the dot operator, must be used to get at the **struct** members in the **matrix** pointer arguments. The dot operator is only correct for **struct**s, and **m1**, **m2**, and **m3** in **madd()** are pointers to **struct**s.

2.12.1. Exercises

1.   An m x n matrix R and an n x k matrix Q can be multiplied to form an m x k matrix product, $P = (RQ)$. We can think of R as a matrix of m row vectors of length n and Q as a matrix of k column vectors of length n. Then the matrix product can be expressed as the dot product of the row vectors in R and the column vectors in Q:

$$P = \begin{bmatrix} \vec{r}_1\vec{q}_1 & & \vec{r}_1\vec{q}_k \\ \cdot & & \cdot \\ \cdot & \cdots & \cdot \\ \cdot & & \cdot \\ \vec{r}_m\vec{q}_1 & & \vec{r}_m\vec{q}_k \end{bmatrix}$$

Write a tool program, **matp**, which takes two matrices conformable for matrix multiplication from **stdin**, calculates their matrix product, and writes the result to **stdout**. Can you use **vdotp()**? Justify your answer.

2.   The transpose of a matrix P, written P', is a matrix in which the rows and columns of P have been interchanged:

$$P' = \begin{bmatrix} p_{11} & & p_{n1} \\ \cdot & & \cdot \\ \cdot & \cdots & \cdot \\ \cdot & & \cdot \\ p_{1n} & & p_{nn} \end{bmatrix}$$

Write a tool program, **transp**, which accepts a matrix as input and writes out its transpose.

2.13. The Algorithm for Gaussian Elimination

The third and final matrix tool presented in this chapter is the program **eqsolv**, which uses Gaussian elimination to solve a set of linear equations like the system:

$$\begin{bmatrix} a_{11} & & a_{1n} \\ \cdot & & \cdot \\ \cdot & \cdots & \cdot \\ \cdot & & \cdot \\ a_{n1} & & a_{nn} \end{bmatrix} \begin{bmatrix} x_1 \\ \cdot \\ \cdot \\ \cdot \\ x_n \end{bmatrix} = \begin{bmatrix} b_1 \\ \cdot \\ \cdot \\ \cdot \\ b_n \end{bmatrix}$$

for the unknown n x 1 vector $\vec{x}$, given the n x n matrix of coefficients A and the n x 1 right-hand-side vector $\vec{b}$.

Written in matrix-vector form, the system becomes:

$$A\,\vec{x} = \vec{b}$$

If A, $\vec{x}$, and $\vec{b}$ were real numbers rather than vectors and matrices, finding a value for x would be simple. We would just multiply both sides of the equation by the reciprocal of A, which would be $\dfrac{1}{A}$. If, however, A were zero, then it would have no reciprocal. In the vector-matrix world, the equivalent of a reciprocal for a matrix is its inverse, while a number without a reciprocal corresponds to a matrix which is singular.

An n x n matrix P is said to be singular if there exists no n x n matrix Q such that:

$$I = P \ Q = Q \ P$$

where I is the n x n identity matrix. Q is called the inverse of P and vice versa. Of course, in the world of real numbers there is only one number with no reciprocal (namely zero), but the number of singular matrices is infinite.

If A is not singular, then both sides of the equation can be multiplied by A^{-1}, which is the inverse of A. $A^{-1} A$ yields the n x n identity matrix, and the identity matrix times $\vec{x}$ gives $\vec{x}$, so the above equation becomes:

$$\vec{x} = A^{-1} \vec{b}$$

giving the solution to the system. If A has no inverse, then the system has no unique solution.

The Gaussian elimination algorithm does not use this method of finding $\vec{x}$, since computing the inverse of A is computationally expensive and completely unnecessary. Instead, it proceeds by doing row operations on the rows of the augmented matrix $[A \mid \vec{b}]$, formed by appending $\vec{b}$ to the right side of A. Legal row operations are:

1. multiplying or dividing a row by a scalar,

2. adding or subtracting one row from another,

3. adding or subtracting a multiple of one row from another.

The object of the row operations is to reduce the matrix A to an upper triangular form while doing all the same row operations on $\vec{b}$ as are done on A. The last element in the vector $\vec{x}$ can then be found by doing a scalar division, since only one nonzero element remains in the last row of A. The other elements of $\vec{x}$ can be found by back substituting in the remaining equations.

In the example system, x_1 can be eliminated from the last n-1 equations by subtracting the factor $\dfrac{a_{i1}}{a_{11}}$ times the first row of the augmented matrix from the ith row of the matrix, provided a_{11} is not zero. x_2 can be eliminated similarly from the last n-2 equations, etc. After all n equations have been treated in this manner, the resulting system is:

$$
\begin{bmatrix} a'_{11} & & a'_{1n} \\ \cdot & & \cdot \\ \cdot & \cdots & \cdot \\ \cdot & & \cdot \\ 0 & & a'_{nn} \end{bmatrix}
\begin{bmatrix} b'_1 \\ \cdot \\ \cdot \\ \cdot \\ b'_n \end{bmatrix}
$$

The primes on the elements of A and $\vec{b}$ indicate that their values have been changed by row operations. x_n can now be found by simply dividing b'_n by a'_{nn}. Back substituting x_n

in the n-1st equation, x_{n-1} can be found, and so forth, through x_1.

Note that, in order to reduce the matrix properly, A must be square. This means that there must be as many equations in the problem as unknowns. Therefore, the conformability criteria for A and $\vec{b}$ are that A is square n x n and $\vec{b}$ must have n elements.

The Gaussian elimination algorithm loops over the columns of A, and, on each iteration of the loop, it selects the largest element in the current column. This element is called the pivot and is divided into the other elements in the column to perform a row operation. In the example above, a_{11} is the pivot. Since other elements in the column are divided by the pivot, the absolute value of the pivot must be at least larger that zero, or, in floating point computer arithmetic, some very small constant close to zero. A larger pivot also helps to reduce numerical errors, so it makes sense to choose the pivot as large as possible.

If the matrix A is singular, then there is no unique solution to the equations. Singularity of A can be detected if the absolute value of the pivot element on a particular loop iteration becomes too small. An appropriate measure of "too small" is:

$$e = ns^{-m}$$

where n is the dimension of A, s is the base of double precision arithmetic on your machine, and m is the number of digits in that base. Most computers use either binary coded decimal (BCD) arithmetic ($s = 10$) or floating point (binary) arithmetic ($s = 2$). The number of digits depends on the computer word size and the number of words used to represent a floating point number.

Assuming we have an n x n matrix A and an n element vector $\vec{b}$, the following pseudo-code returns in the vector $\vec{x}$ a solution to the system or an indication that the system cannot be solved:

Let **err** be the lower bound for an acceptable pivot;

/*reduce A[][] to upper triangular form*/

Loop over the columns of **A[][]**, letting **k** be the column index:
{

    pivot = A[0][k];
    maxr = 0;

    Loop over the rows of **A[][]** to find **pivot**, letting **i** be the row index:
    {
        if(ABS(pivot) < ABS(A[i][k]))
        {
            pivot = A[i][k];
            maxr = i;
        }
    }

    /*check if the pivot is too small*/

    if(ABS(pivot) <= err)
        return(ERR);

    Swap rows **k** and **maxr** in **A[][]** and **b[]**, putting the pivot element on the diagonal;

    Loop over the rows of **A[][]** from row **k+1** to **n-1**, letting **i** be the row index:
    {
        A[i][k] = A[i][k] / A[k][k];

        Loop over the elements in row **i** from column **k+1** to **n-1**, letting the column index be **j**:
        {
            A[i][j] = A[i][j] - A[i][k] * A[k][j];
        }

        b[i] = b[i] - A[i][k] * b[k];
    }
}

/*back substitute to find the solution*/

Loop over the rows in reverse starting with **n-1**, letting **i** be the row index:
{

    multiplier = 0;

    Loop over the elements in row **i** starting with column **i+1** to **n-1**, letting **j** be column index:
    {

        multiplier = multiplier + A[i][j] * x[j];

    }

    multiplier = b[i] - multiplier;

    if(the ith diagonal element of **A[][]** is larger than **err)**
    {

        The ith element of **x[]** is calculated by dividing **multiplier** by **A[i][i];**

    }
    else if(multiplier < err)
    {

        The ith element of **x[]** is 0.0;

    }

    /*error in calculation*/

    else
        return(ERR);
}

return(OK);

2.14. A C Implementation

The C implementation of Gaussian elimination follows straightforwardly from the pseudo-code definition. In place of the matrix A a **matrix** pointer **A** is passed to the function. The lower bound on the pivot, **err**, is calculated using the C mathematical function **pow()**. The function **pow()** takes two arguments, a **double** base and a **double** exponent, and returns the base to the exponent power:

    double a,base,exponent;
    double pow();

    a = pow(base,exponent);

Although **leqsolv()** is rather long as C functions go, the loop structure of the code follows the pseudo-code outline closely. Here is the code:

```
/*
**********************************
leqsolv-solve the linear system Ax = b using
        Gaussian elimination with partial
        pivoting.
**********************************
*/

#define ABS(a)              ( (a) >= 0 ? (a):-(a))

int leqsolv(n,A,b,x)

  int n;
  matrix *A;
  double b[],x[];

{
  int i,j,k,maxr;
  double err,pivot,*rows[MAXVEC],m,*rowswap;
  double pow();

    err = (double)n *
        pow(MACHINE_BASE,-MACHINE_DIGITS);

/*
initialize row pointer array
*/

    for( i = 0; i < n; i++)
      rows[i] = A->m[i];

/*
loop over columns of A->m
*/

    for( k = 0; k < n-1; k++)
    {
      pivot = err;
      maxr = 0;
```

```
/*
find pivot element
*/

    for( i = k; i < n; i++)
    {
      if( ABS(pivot) <= ABS(rows[i][k]) )
      {
        pivot = rows[i][k];
        maxr = i;
      }
    }

/*
check if pivot is too small
*/

    if( ABS(pivot) <= err)
      return(ERR);

/*
swap rows k and maxr
*/

    rowswap = rows[k];
    rows[k] = rows[maxr];
    rows[maxr] = rowswap;

/*
swap right hand sides
*/

    m = b[k];
    b[k] = b[maxr];
    b[maxr] = m;

/*
loop over rows
*/
```

```
      for( i = k+1; i < n; i++)
      {
```

```
/*
compute multiplier
*/
```

```
        m = rows[i][k] = rows[i][k]/rows[k][k];
```

```
/*
compute new row elements
*/
```

```
        for( j = k+1; j < n; j++)
          rows[i][j] = rows[i][j] - m * rows[k][j];
```

```
/*
compute right hand side sum
*/
```

```
        b[i] = b[i] - m * b[k];
      }
    }
```

```
/*
loop over rows in reverse
*/
```

```
    for( i = n-1; i >= 0; i--)
    {
      m = 0.0;
```

```
/*
substitute in right hand side
*/
```

```
        for( j = i+1; j < n; j++)
          m += rows[i][j]*x[j];
```

```
        m = b[i] - m;
```

```
/*
check for zero pivot
*/
```

```
            if( ABS(rows[i][i]) > err)
              x[i] = m/rows[i][i];

/*
zero right hand side
*/

            else if( ABS(m) <= err)
              x[i] = 0.0;

/*
error in calculation
*/

          else
            return(ERR);
        }

        return(OK);
}
/*
end of leqsolv
*/
```

Several places in the code require testing the absolute value of a number, to see if it is too small. We could have used the function **abs**() from Chapter 1, but instead the macro **ABS**() is used to perform the calculation. The advantage of using a macro over a function call in this case is that the arguments of **ABS**() can have any type (**double, float, int,** or **char**) and only one macro is needed for all types. The macro could potentially be faster, since the C preprocessor will expand the macro as in-line code. If the macro argument is a complicated expression, however, using a macro could well be slower than a function call, since the expression will be evaluated twice in the macro, but only once when the function is called. As in Chapter 1, both the argument and the macro code itself were protected from misinterpretation when substituted into expressions by surrounding them with parentheses. Since **leqsolv**() only has variables as arguments to the **ABS**(), double evaluation isn't a problem.

The **ABS**() macro illustrates another interesting kind of operator available in C: the conditional (**?**) operator. The syntax for a conditional expression is:

expression1 **?** expression2 **:** expression3

When a conditional expression is encountered, expression1 is evaluated first. If the value of expression1 is not zero, then the value of the entire expression results from evaluating expression2, otherwise expression3 is evaluated and that is the result.

The conditional operator has a very low precedence, so conditional expressions should be enclosed in parentheses to ensure that they are evaluated as a unit. A conditional expression can be used wherever a normal arithmetic or logical expression can, and is thus a kind of "computational **if** expression." It can save a long chain of **if-else** statements whose only purpose is to do a numerical computation that differs depending on the value of a variable. Overuse of the conditional expression should be avoided, however, since it tends to make code cryptic and difficult to read.

leqsolv() uses pointers to speed up and simplify what could otherwise be a time-consuming and complex operation, namely exchanging the rows of the matrix to put the pivot on the diagonal. The Gaussian elimination algorithm calls for exchanging the rows of **A->m[][]** so that the pivot element is on the diagonal. Copying rows requires extra storage in the form of a vector to hold one row while the other is being copied into it, and could require up to n^2 assignments for a matrix with n rows and columns. If, however, pointers to rows in the matrix are swapped instead, only n assignments are needed. The code for **leqsolv()** takes pointers to the rows of **A->m[][]** before the calculation begins. **A->m[i]** can be thought of as a pointer to the ith row of **A->m[][]**. These pointers are assigned to the **double *** array **rows**. All subsequent calculations are done with the pointer array rather than with **A->m[][]** itself. Note that we cannot use the same trick with the columns, because columns are not stored contiguously in memory.

2.15. The Rest of eqsolv

The **main()** and the routine for returning a matrix and vector conformable for Gaussian elimination differ very little from **dotp** and **addm**. The main routine has exactly the same structure, except the routines **getvecs()** from **dotp** and **getmats()** from **addm** are replaced by **getmatvec()**, which returns a matrix and vector conformable for Gaussian elimination. **leqsolv()** is called to do the calculation, and **fputv()** puts out the resulting vector. The return status of **leqsolv()** is checked, and an error message is issued if **A->m[][]** is singular. The matrix is also written out in case of error using a routine **fputm()**. The code is a straightforward extension of that from **dotp** and **matp**:

```
/*
*******************************
eqsolv-read matrix, A, and vector, b,
        from standard input, calculate
        solution to the matrix equation
        Ax = b, and write out the result.
*******************************
*/

#include <stdio.h>
#include "c.h"
```

```
            main()

         {
           int n;
           int getmats();
           matrix A;
           double b[MAXVEC],x[MAXVEC];

             while( (n = getmatvec(&A,b)) != EOF)
             {
               if( leqsolv(n,&A,b,x) == ERR)
               {
                 fprintf(stderr,"eqsolv:matrix is singular.\n");
                 fputm(stderr,&A);
               }
               else
                 fputv(stdout,n,x,COL);
             }

             exit(0);
         }
         /*
         end of main
         */
```

The function **fputv()** writes a vector out to a file passed in as a **FILE** pointer. In addition to the **FILE** pointer, the length of the vector, and the **double** vector itself, **fputv()** takes an integer parameter indicating whether the vector is to be written out as a row vector or as a column vector. If the parameter has the value **COL**, then the vector is written out as a column vector; otherwise, a row vector is assumed. The constants **ROW** and **COL**, for indicating row and column vectors, respectively, are defined in **c.h**. In most cases, the vector can be written simply as a row vector. Sometimes, however, as in the case of **x** in **eqsolv**, whether the vector is a row or column vector does matter. Mathematically, **x** must be a column vector, or an n x 1 matrix; otherwise, A and $\vec{x}$ are not conformable for matrix multiplication. While this probably won't make a difference for people viewing the answer, it might if some other numerical tool uses the result generated by **eqsolv** to do some further calculations. Technically, the same is true of $\vec{b}$ on input, but row vectors are faster for people to input so both are accepted. More about this in the next chapter.

getmatvec() checks both that **A** is a square matrix and that the column dimension of **A** and the number of elements in **b** are equal. Both are prerequisites for Gaussian elimination to proceed. Other than these error checks, **getmatvec()** is not very much different from **getvecs()** or **getmats()**, and the source is therefore not presented.

2.15.1. Exercises

1. Investigate algorithms for finding the inverse of a matrix and write a matrix inversion tool called **inv** which inverts matrices typed as input on **stdin**.

2.16. Errors and Numerical Instability

The implementation of Gaussian elimination in the previous section does not address several problems which may occur during computation. In particular, matrices A and vectors b with elements that differ by several orders of magnitude are not well solved by this implementation. Scaling the matrix elements so that they are all about the same order of magnitude can help, although this is not often possible. Round-off errors are another possible source of inaccuracy. The pivoting method chosen here should help to reduce round-off errors in most problems. Methods for estimating round-off error and improving the calculated answer based on the estimate are discussed in the chapter references.

However, disregarding round-off errors, even a matrix and vector in which the order of magnitude of the elements is about the same can lead to problems. Consider the following example:

$$A = \begin{bmatrix} 1.2969 & 0.8648 \\ 0.2161 & 0.1441 \end{bmatrix}$$
$$\vec{b} = (0.8642, 0.1440)$$

If the estimated solution is given as:

$$\vec{x} = (0.9911, -0.4870)$$

then the difference between the actual value of $\vec{b}$ and the value calculated by multiplying A and $\vec{x}$, $\vec{b} - A\vec{x}$, is:

$$(-10^{-8}, 10^{-8})$$

Even though this seems to indicate that $\vec{x}$ is correct, the exact solution is actually:

$$\vec{x} = (2, -2)$$

as can be seen by solving $\vec{b} - A\vec{x}$ with $\vec{x} = (2, -2)$.

The problem occurs because, during one of the intermediate steps in the elimination, the difference $a_{ij} - \dfrac{a_{ik}}{a_{kk}} a_{kj}$ is on the order of 10^{-8}. In order to maintain accuracy, the elements in A and $\vec{b}$ would need to be specified to a precision better than 10^{-8}. This kind of behavior is known in the numerical literature as ill-conditioning. Methods exist for determining whether a matrix is ill-conditioned, and many commercial Gaussian elimination routines will return an indication of how well conditioned a problem is, if requested. In the references at the end of the chapter, ill-conditioning is discussed in more detail.

Estimating round-off errors and determining whether a problem is ill-conditioned have associated computational costs. For the initial phases of a numerical investigation, error analysis might not be necessary; however, as the results of the investigation are refined, accuracy becomes important. In the final analysis, the results of a numerical investigation are only as worthwhile as their accuracy. If your problem demands an estimate of the size of round-off errors, or you have reason to believe that the problem may be ill-conditioned, the simple Gaussian elimination routine given here should be modified to do some error analysis.

2.17. Organizing Program Development

The organization of the files in which code for the tool programs is contained is almost as important as structuring the code itself. The hierarchical nature of the UNIX file system can be used to segregate the files for programs with similar functionality into particular directories. The UNIX file system starts at the root, designated /, and any directory or file within the file system can be designated with a pathname. A UNIX pathname consists of the directories in the path from the root separated by slashes (/). An example is **/users/jones/tools/vec-mat**. The last name in a pathname can be either a directory or a file. The directory above the current one is called the parent directory and can be abbreviated **..**, so a file or directory in the **tools** directory could be accessed using the pathname **../thing**, if the current directory were **vec-mat**. The current directory can also be abbreviated **.**, however, files and directories in the current directory can be addressed simply by their name. If **vfile** is a file in the current directory, then it can be referred to either as **vfile** or **./vfile**.

Users are generally given a login directory, something like **/users/<username>**, with the exact pathname depending on custom and the system administrator of the local system. When you log in, this directory becomes your current working directory, and all commands, to create or delete files or execute programs, are relative to this directory. Putting all your files into this directory is one way to guarantee that you'll never be able to find anything. Instead, subdirectories for programs having particular functionality can be created using the UNIX file system **mkdir** command:

$ mkdir tools

Typing this command to the UNIX shell will create a directory where files or further subdirectories can be created.

Once a directory has been made, you can change to it using the command **cd**:

$ cd tools

Now, all file and directory creation is relative to the directory **tools** under the login directory.

An even finer distinction may be desirable if a particular program or program group has lots of source files. Within the directory for a particular program or program group, the files **src**, **obj**, and **bin** can be used to hold the source, object (compiled but as yet un-linked), and executable files. This nicely separates the files so that a directory listing (generated using the UNIX **ls** command) doesn't run to multiple screenfuls of information.

Once a directory structure has been created for a program or program group, some thought should be given on how to organize the files which will contain the source. On the one hand, lumping everything into one file causes recompilations to take a long time, since a small change in a single routine means the entire file must be recompiled. On the other, putting each function into a separate file makes keeping track of all the modules needed at link time very difficult. A happy medium is to group routines by functional similarity.

For the three programs presented in this chapter, the **main**() function and the functions fetching and verifying the input data were grouped into one module for each program: **dotp.c** for **dotp**, **matp.c** for **matp**, and **eqsolv.c** for **eqsolv**. The routines for doing vector operations, like **vdotp**(), were put into a file called **vector.c**, while the matrix addition routine was put into **matrix.c**, where routines which operate on matrices are stored. **leqsolv**() is complicated enough to merit its own file, **leqsolv.c**. Finally, two additional files are handy for storing functions which are used throughout the book, and for defined constants like **MAXVEC**. These are **c.c** and **c.h**. Functions in **c.c** can be considered a supplement to the C library. Although we haven't encountered any of these functions in this chapter, the function **getword**(), which is described in Chapter 3, is a candidate. Every source file will contain an **#include** at the top for **c.h**, since the constants in **c.h** are useful in a wide variety of programs. Some constants contained in **c.h** are described in Section 2.1. Practically every program in the book makes use of things from **c.c** and **c.h** in some way.

Although the code for the lower level functions that get/put vectors and matrices to/from files has not yet been presented, the code for the main routines, routines which verify the input data, and the numerical routines can be typed into files using one of the UNIX text editors. The calls to **pgetv**(), **pgetm**(), **fputv**(), and **fputm**() can be replaced by stubs. Stubs are pieces of code which, in some way, simulate what the actual function should do. At the very simplest, they may print out a message that they have been called and return. The control flow of the program can be checked in this manner, to be sure each function is being called correctly. More complete testing can be done by making each stub return some sample vector or matrix which is directly wired into the code. For example, a stub for **pgetv**() might look like:

```
/*
***************************
pgetv-stub to test code for dotp.
***************************
*/

int pgetv(fin,fout,v)

  FILE *fin,*fout;
  double v[];

{

    v[0] = 1.0;
    v[1] = 2.0;
    v[2] = 34.0;

    fprintf(fout,"pgetv:returning v as 1.0 2.0 34.0.\n");

    return(3);
}
/*
end of pgetv
*/
```

Similarly, the vector and matrix output functions can output each element using **printf()** by including the number of elements in the test vector or matrix as a constant. The stubs can be put into a file called **dummy.c**, compiled, and linked with the rest of the code to test the numerical algorithms. Later, the stubs can be replaced by the functions which do the actual vector-matrix i/o.

Compiling each of the modules separately is almost as simple as compiling and linking in one step, which was discussed in Chapter 1. The command line for the UNIX shell:

> **$ cc -c** <filename>**.c**

causes the C compiler to compile the file with name <filename>**.c** and and put the compiled file into a relocatable object module <filename>**.o**.

When the object code is contained in several different modules, the modules can be linked together to produce an executable program in the following manner:

> **$ cc** <file1>**.o** <file2>**.o ...** <filen>**.o -o** <program name>

The files <file1>**.o** through <filen>**.o** contain the object modules to be linked together.

When the files in the command line contain compiled code, **cc** starts the linker with some standard arguments and libraries, to link the object modules together into an executable module. The executable will be left in <program name> and can be executed by simply typing <program name> at the terminal. On the UNIX system, the linker is called **ld** and can be invoked without going through **cc**, though a more complex set of options and library names must be specified if so.

In the case of **eqsolv**, an additional module is needed, since reference is made in **leqsolv**() to the math library function **pow**(). The location of the math library is known to the linker. If a math library function is used, the argument **-lm** included on the command line after the list of object modules tells the linker to get the math library and search it for the functions used in the program. The command line for linking **eqsolv** is:

$$\text{\$ cc eqsolv.o matrix.o vector.o dummy.o -lm -o eqsolv}$$

The location of the **-lm** argument is important, since the linker searches object and library modules in the order they appear on the command line, and will not resolve references using the library if the library argument comes before the object file containing the function call. This is also true for the object modules themselves, so the main module must appear first, and modules containing calling functions must appear before the modules containing the functions they call.

The UNIX system has a number of other software tools for supporting program development. The steps involved in making an executable can be collected together into a file and the **make** program can be used to build the executable, instead of having to retype each individual step every time the executable must be built. **make** is particularly useful when a program is built out of many source modules, since it figures out which files have been modified since the last time the executable was made and only remakes those which were changed. Another valuable software tool is **grep**, a program which searches through source files for strings. If you want to find out all the places where a particular variable is used, for example, **grep** can search through the files containing the source and list out the files and source lines where the variable occurs. For more information on these and other UNIX software tools, check the *The UNIX Programmer's Manual.*

2.18. References

Some good books covering all aspects of vector-matrix computation are:

1. *Calculus and Analytical Geometry*, by George B. Thomas, Addison-Wesley, Reading, MA, 832 pp., 1972.

    Discusses the analytical aspects of vector-matrix computation.

2. *Numerical Methods*, by Germund Dahlquist, Ake Bjoerck, and Ned Anderson, Prentice-Hall, Englewood Cliffs, NJ, 576 pp., 1974.

    Contains good discussions on the topics of numerical error and accuracy, as well as a large section on numerical linear algebra. The example of an ill-conditioned problem in Section 2.16 was taken from here.

3. *Matrix Computations and Mathematical Software*, by John Rice, McGraw-Hill, New York, NY, 256 pp., 1981.

    Much good and very specific information on solving linear equations. The description of Gaussian elimination in Section 2.13 follows from this source. Also talks about commercially available software.

4. *The Engineering of Numerical Software*, by Webb Miller, Prentice-Hall, Englewood Cliffs, NJ, 176 pp., 1984.

    A very appropriate treatment of the software engineering aspects of writing numerical software, from the FORTRAN perspective. Contains a chapter on solving linear equations.

3

EXCHANGING VECTOR
AND
MATRIX DATA

Although the upper-level routines for **dotp**, **addm**, and **eqsolv** were developed in the last chapter, the implementation of i/o routines to communicate vector and matrix data between peripheral devices like the user's terminal and the programs was not specified. The function-calling interface for the i/o routines was discussed in detail, as was the general nature of what the functions needed to do. The input functions which will be developed in this chapter must guarantee the return of a correct vector or matrix, or the constant **EOF** as the value of the function if the end of file was reached. The output functions simply need to write their parameter data in the proper format to the output file associated with their **FILE** pointer parameter.

In addition to these specifications, a more vague design requirement is that all three matrix tools presented in Chapter 2, and potentially more to come, should be able to freely exchange numerical data between themselves without the need for extensive interconversion programs. All tools which operate on vector and matrix data should be able to use the same set of i/o routines. Two features of the UNIX operating system, pipes and i/o redirection, provide the supporting foundation for this goal. Upon that, a simplified vector-matrix exchange protocol is developed in this chapter, along with the routines to implement it.

3.1. Exchanging Data with Pipes and I/O Redirection

Many operating systems have no ready-made interface allowing programs to exchange data with each other. If the programmer wants two programs to exchange data, an intermediate file must be used, and the programs must take care of making the proper connections to that file. With the UNIX system, however, programs can be run concurrently, and an interface, called a pipe, is provided for data exchange.

A pipe can be used to direct the output of one program to the input of another without the need of a intermediate file. Consider the following line, typed to the UNIX shell:

$ prog1 | prog2

When this line is interpreted by the UNIX shell, **prog1** and **prog2** are started simultaneously and run concurrently. If **prog1** writes byte data to its standard output, **prog2** can read that data from its standard input. Thus data can be passed easily between programs, without the need of an intermediate file. The resemblance to a water or gas pipeline, with the byte data exchanged between programs being the fluid, led to the name. Programs which read data from a pipe, transform that data in some manner, and write the data back onto the pipe are often called filters.

This same philosophy of easy data transfer underlies the method of redirecting input or output to a file, if the program uses **stdin** and **stdout** for i/o. As discussed briefly in the previous two chapters, all C programs normally begin with three open files: standard in (**stdin**), standard out (**stdout**), and standard error (**stderr**). **stdout** and **stderr** are associated with the terminal screen, while **stdin** is associated with the terminal keyboard.

Both **stdin** and **stdout** can be redirected to a file by using the < and > operators, a process called input or output redirection. For example, if **prog1** reads data from **stdin** and writes to **stdout**, the following line typed to the UNIX shell will cause **prog1** to read from **file1** and write to **file2**:

$ prog1 < file1 > file2

No change in the code for **prog1** is needed, nor do any special command files or operating system commands need to be run to accomplish redirection.

Many of the UNIX text-processing commands were designed to be used in pipes. The program **cat** reads a series of files, whose names are given as command line arguments, and writes them to **stdout**. **wc**, another UNIX filter, takes text data from **stdin**, counts the number of characters, words, and lines, and writes the resulting count to **stdout**. Using the two in a pipe, the command

$ cat file1 file2 | wc

counts the number of characters, words, and lines in **file1** and **file2**. Similar but more powerful text-processing can be achieved by using other UNIX filters.

If all our numerical tools read their input data from **stdin** and write to **stdout**, then two tool programs can be used together in a pipe to perform a function which otherwise would have required a whole new program to be written. Since all of the vector-matrix tools presented in the last chapter were developed with this objective in mind, it should be possible to perform multiple vector-matrix operations by simply combining the tools in pipes. The result is a large saving in time and effort which would otherwise have gone towards writing and debugging the extra code.

For example, if A is a nonsingular n x n matrix, and $\vec{b}$ and $\vec{c}$ are n element column vectors, the matrix operations:

$$A^{-1}\vec{b} + \vec{c}$$

can be performed in the single command:

$ eqsolv | addm

since $A^{-1}\vec{b}$ is the solution to the equation $A\vec{x} = \vec{b}$. Other, similar matrix operations can be done using pipes and the matrix tools developed in Chapter 2.

3.1.1. Exercises

1. Give an example of a matrix equation which cannot be solved using a single command line. Can it be solved using i/o redirection in addition?

3.2. Adapting Pipes to Vector and Matrix Data

The pipe interface as it stands is sufficient if the data being exchanged by programs is simply characters. Programs need only read characters from **stdin**, process them, and write characters to **stdout**. The character stream will be passed from program to program along the pipe.

Most UNIX word processing utilities operate on text lines rather than on individual characters. These programs read lines of text from **stdin**, process the lines, and write the results to **stdout**. The additional structure of text lines is imposed on a stream of character data by simply giving one character, the newline ('\n'), a special function. A newline character is used to mark the end of a text line. Programs can differentiate between text lines by looking for newline characters.

In the case of vectors and matrices, however, the data contains considerably more structure. At the lowest level, numerical data will consist of C **double**s. These correspond to the characters which are the smallest units operated upon by word processing programs. Unfortunately, a single **double** written into a stream of characters on a pipe is made up of one or more characters. Some character is needed to delimit one **double** from another.

At the next higher level, vectors correspond to the text lines exchanged by word processing utilities. A single vector contains one or more **double**s and must be delimited from the next vector by some special character. An additional level of structure exists above the vector.

Since a matrix consists of several vectors, some indicator is needed to distinguish a matrix from a stream of vectors. The corresponding level for text processing utilities would be the paragraph, although the UNIX word processing utilities don't recognize any text structure higher than the line.

There are two types of vectors which need to be distinguished. If A is an n x n matrix and $\vec{b}$ is an n element vector, then the matrix product $A\vec{b}$ is an n x 1 matrix, or a column vector. Row vectors, consisting of a single text line of **double**s, can also occur. Programs like **dotp** don't care whether the vectors they process are column or row vectors, and the interface protocol should reflect this. On the other hand, some matrix tools may care whether one of their operands or the result of their calculations is a row or a column vector; **eqsolv** is an example. The result of solving a linear system is an n x 1 matrix, or a column vector. Technically, the input right hand side vector $\vec{b}$ should also be a column vector, but the format doesn't really matter for the calculations (as it would for matrix multiplication), so either format is accepted.

The differentiation between row and column vectors is made in **fputv**(). A parameter is passed to **fputv**() indicating whether a vector should be written as a row or column vector. This was mentioned briefly during the discussion of **eqsolv** in the last chapter. In addition, some provision must be made in **pgetv**() to accept both row and column vectors, since a tool which doesn't care about the shape of its vector input must be able to accept both.

If we adopt the convention that one or more **double**s delimited on a text line by "white space" (i.e., spaces and tabs) constitute a row vector or matrix row and a series of text lines containing row vectors constitutes a matrix, then we are left with the need to differentiate a series of row vectors from a matrix. A single blank line will be used to separate a vector or matrix from the next vector or matrix in the input stream. In addition, if a single vector or matrix row is too long to fit on one line, the backslash character (\) as the last character on the line can be used to continue numerical input on the next line.

A vector or matrix row can be treated exactly like a line of text and the elements can be read or written using the formatted i/o facilities of C. An example of a row vector is:

    5.1   2.3   4.7   89e-2
    <blank line>

White space at the beginning of a line and at the end of a line will be ignored. A text line which would not be acceptable as a vector is:

    1.2   q   7a   3.8
    <blank line>

since **q** and **7a** are not acceptable C **double**s. An example of a column vector is:

4.3
2.1
5
<blank line>

On input, a column vector can be treated like an n x 1 matrix, if the vector-matrix tool absolutely requires a column vector as one of its operands.

Notice that this data exchange protocol does not require a user to tell the program beforehand how many elements a vector or matrix will have. The user need only signify when data entry is complete, and the routines that fetch the data will do the rest, simplifying data entry.

From the description of the data exchange interface given so far, an interconversion program would be needed in only one case. If we wanted to construct a pipe in which the first member operated on matrices but the second operated upon the rows of the matrices as if they were row vectors, we would need an interconversion program between the two tools to reformat the matrices as row vectors and write them out. Likewise, if a tool which wrote out vectors were followed by a tool which treated those vectors as rows of a matrix, another interconversion program would be needed. Fortunately, combinations of the common matrix and vector operations performing these kinds of operations are rare.

So far, we've talked about how individual programs get their input. When two programs are put together in a pipe, however, all input goes through the first program in the pipe before reaching the second. Unlike the UNIX text-processing filters, there may be times when the data read by a numerical tool on a pipe may need to be protected from processing. This point was briefly discussed in the previous chapter, when the function call to **pgetv()** was introduced.

For example, suppose we want to do the operation in the last section, $A^{-1}\vec{b} + \vec{c}$. If no provision is made to protect $\vec{c}$ from processing by **eqsolv**, then $\vec{c}$ will be "caught" by **eqsolv** before it reaches **addm**, and **eqsolv** will try to use it in a computation. Only the first program in the pipe is connected to the keyboard, so vectors which are only used by later members of the pipe must pass untouched through all preceding programs before reaching their goal. **eqsolv** must be told to pass $\vec{c}$ along without further processing, and not to try to use it in computation.

Graphically, the easiest way for a user to enter data bound for a particular program is to have the data entered at approximately the same position on the screen as the program name in the command which started the pipe. The vertical bar pipe symbol separates data bound for the different programs on the pipe. In the above example, if A is the matrix:

$$\begin{bmatrix} 2 & 7 \\ 4 & 19 \end{bmatrix}$$

and $\vec{b}$ and $\vec{c}$ are the vectors:

$$\vec{b} = (6, 13)$$

$$\vec{c} = (21, 1)$$

the data for the computation is entered as follows:

```
2    7   |   21
4    19  |   1
         |
6    13
```
<blank line>

The vertical bar on the line after the matrix is needed as a signal to **addm** that the vector $\vec{c}$ has been entered, since vectors require a single blank line after the line containing the vector body. $\vec{c}$ must be entered as a column vector in this case, because the output of **eqsolv** is a column vector and the two must be conformable for addition. The resulting output of the two-program pipe containing **addm** and **eqsolv** is:

```
23.3
1.2
```

This input format bears a close resemblance to the corresponding vector and matrix operations written on paper, and it provides a comfortable interface to users. Of course, the basic line-oriented nature of the tools' i/o makes it difficult to duplicate that layout exactly. The exercises at the end of this section deal with this and other problems in more detail.

3.2.1. Exercises

1. If, in addition to the tools **dotp**, **addm**, and **eqsolv**, you had available a tool, **matp**, which calculated the matrix product, give one or more command lines using **dotp**, **addm**, **matp**, and **eqsolv** that perform the following operations. As in the example, upper case indicates n x n matrices and lower case indicates n element vectors.

    i. $((A \ A) \ \vec{b}) \ \vec{c}$

    ii. $\vec{c} + (A \ \vec{b})$

    iii. $\vec{c} + \vec{b}$

    iv. $(A^{-1} \vec{b}) \ \vec{c}$

2. What would the input sequence corresponding to the above command lines look like? What would the output be?

3. Say an additional tool program, **sadd**, is available, which reads two scalars (i.e., vectors with just one element) from **stdin**, adds them together, and writes the result to **stdout**. What modifications, if any, would have to be made in the interface protocol described to accommodate the new tool? Give an example in which **matp**, **dotp**, and **sadd** are used. What would the command line look like? The input sequence? The output?

4.   What kind of information about vectors and matrices would have to be exchanged
      on a pipe to be able simply to type in vectors and matrices as they are written on
      paper and have them get to the proper program? To avoid having to differentiate
      between vectors and rows in a matrix?

3.3. The Output Functions fputv() and fputm()

The simplest part of the i/o interface is the output routines **fputv()** and **fputm()**. Both
have the task of converting their respective data structures into character form and writing the
result to an output file specified as a parameter of type **FILE ***. In addition, an empty line
must be written out after each vector or matrix, to indicate that the data is complete.

The code for **fputv()** is straightforward enough not to require pseudo-code:

```
/*
*****************************
fputv-put out vector v to file fout.
*****************************
*/

int fputv(fout,len,v,isa_colvec)

  FILE *fout;
  int len,isa_colvec;
  double v[];

{
  int i;

    if( len > MAXVEC)
      return(ERR);
/*
 put out elements, on separate row if a column vector
*/

    for( i=0; i < len; i++)
    {
     fprintf(fout,"%6.16g ",v[i]);

     if( isa_colvec == COL)
       putc('\n',fout);
    }
```

```
/*
put out a newline to terminate elements, if a row vector
*/

   if( isa_colvec != COL)
     putc('\n',fout);

/*
 put out empty line to terminate vector
*/

   putc('\n',fout);

   return(OK);
}
/*
end of fputv
*/
```

The parameter **isa_colvec** should be set by the calling routine to **COL**, if the vector is a column vector. The default is to treat **v[]** as a row vector. After a check is made to be sure the length of the vector doesn't exceed the maximum, **fputv()** loops over all elements in the vector, starting with the first one, and writes them to the output file. If the vector is a column vector, a newline character is written out after each element, positioning each element on a separate line. The terminating newline is written after the last element, so the vector is identifiable by other programs on a pipe as a vector (or matrix with one row or column).

The precision, or number of digits after the decimal point, for numbers printed to the output file in **fputv()** must be set high enough to avoid truncating digits from **doubles** exchanged between programs. Remember that two numerical tools communicating through a pipe first convert the data to character form. If the precision in **fputv()** is below the machine precision, digits will be truncated from the **doubles** in the vector and a loss of precision, a kind of round-off error, will occur between programs. For this reason, the precision in the format string of the **printf()** should be set as high as the machine's floating point precision will allow.

The code for **fputm()** is similar, except two loops are needed, an outer one over rows and an inner one over columns, and, of course, the checks for a column vector are not needed. The implementation of **fputm()** is left as an exercise.

3.3.1. Exercises

1.   Write **fputm()** using **fputv()** as a base.

2.   Do you think a column vector should be put out by a separate function? If so, implement a function which does so. When would a flag like the one to **fputv()** be a better idea than a separate function?

3.4. Error Checking

Error checking is a particularly troublesome part of writing any program. On the one hand, overly redundant checking tends to clutter up the code and can even slow down calculations. On the other, skimpy checking runs the risk of letting unacceptable data slip through or having the program blow up because an out-of-bounds array or pointer reference occurred.

Some (but very few) errors are more efficiently caught and reported by the hardware or operating system. An example is the floating overflow check discussed in Section 2.2. The conclusion there was that the benefits of adding code to catch a floating overflow would not have exceeded the disadvantages of slowing down the dot product calculation. Since the hardware floating overflow trap gives no information, relative to the source code, about the cause of the error, a floating overflow in **vdotp()** during the course of a complex calculation may be extremely difficult to trace. However, that risk balances off against the large majority of cases where a software floating overflow check would cause a slowdown in a complex calculation which involved the dot product.

In Chapter 2 and the previous sections of this chapter, the error checking for particular functions was described as they were presented, on a case by case basis. Errors in numerical programs tend to fall into two general categories:

1.   Attempts to use a particular algorithm on data for which it is unsuited or on which it cannot be used. Errors during execution on a particular data set also fall into this category. An example would be an attempt to do Gaussian elimination on a singular matrix.

2.   Errors arising from the computer implementation of an algorithm. An attempt to read more than **MAXVEC** elements into a vector in **dotp** is an example.

The first type of error should always be caught and informatively reported whenever convenient. If a program or function can deal only with data of a certain type, then often the most effective course is to prescreen the data as it is being input, and write out error messages if invalid data is found. This course was used in **getvecs()**, **getmats()**, and **getmatvec()**. In these functions, a correct vector or matrix could be entered by the user, but because the number of elements in the vector or dimension of the matrix did not match previously entered data, the vector or matrix would not be appropriate input data for the algorithm. **getvecs()**, **getmats()**, and **getmatvec()** solved the problem by checking the data, and writing an error message out informing the user what the problem is and in which program the problem occurred.

On the other hand, some errors only become apparent while a particular numerical operation is proceeding. A singular matrix could be detected in **leqsolv()** only while the solution to the system was being sought through Gaussian elimination. Since **leqsolv()** could potentially be called by a number of programs, and, since singularity is the only error condition which could develop (barring oversized matrices and vectors), the simplest course is to return an error indication to the calling routine and let the calling routine report it.

Errors of the second type should be caught and reported as early as possible, and should not be allowed to propagate. However, redundant error checking should be avoided. In **dotp**, **addm**, and **eqsolv**, the size of matrices and vectors generated during the program is strictly determined by the size of the input data. If **pgetv()** and **pgetm()** report attempts to read over-sized vectors and matrices, the other routines which do calculations with vectors and matrices don't have to. On the output end, the error checks in **fputv()** and **fputm()** make sure that over-sized matrices and vectors generated by one program don't propagate through a pipe. /xe "error checking"

3.4.1. Exercises

1. Can you think of a case where checking for oversized vectors or matrices should be done after the input functions are called?

3.5. pgetv()

The task of **pgetv()** is to get a syntactically correct vector from a file whose **FILE** pointer is passed as the first parameter in the parameter list, and return that vector to the calling routine. The vector may have been entered as a column or row vector, and the **FILE** pointer may have been opened to a pipe, to a file on disc (using **fopen()**) or may simply be **stdin**. When **pgetv()** returns, either the end of file was encountered, in which case the constant **EOF** is returned as the value of the function, or a syntactically correct **double** vector was fetched and the return value of the function is the length of the vector. Any lines containing the pipe symbol must be truncated, and all data after the pipe symbol must be written to the output **FILE** pointer passed as the second parameter. Errors in the input must be reported by **pgetv()** and routines called by it.

This is a formidable task for a single function, and **pgetv()** accomplishes it by delegating subtasks to lower-level functions. The main body of the routine is a **while** loop which fetches rows of **double**s from the input file until a syntactically correct vector is obtained, or the end of file is reached. In pseudo-code, this becomes:

```
finished = NO;

while( finished != YES )
{
    Get a row of doubles from the input;

    if( end of file reached )
        return(EOF);

    else if( first time through the loop )
    {
        Set the vector length to the number of doubles returned;

        if( more than one double returned )
        {
            This is a row vector and no doubles should be returned next time;
```

```
            }
            else
            {
                This could be a column vector and a maximum of one double can
                be returned next time;
            }
        }
        else if( nothing returned and a vector was collected)
            finished = YES;

        else
        {
            Increment the vector length by the number elements returned;
        }
    }

    return( vector length);
```

An integer variable, **finished** has the value **NO** until a syntactically correct vector has been collected. The task of fetching the row of **double**s from the input, converting from character form to binary, and taking care of the pipe protocol and continuation of input over the end of a line is handled by **getdoubles()**. The two **FILE *** parameters to **pgetv()** are passed along to **getdoubles()**, as is the **double** vector parameter. In addition, **getdoubles()** takes an **int** parameter **bound**, which is an upper bound on the number of elements that **getdoubles()** can read in. By resetting **bound** when a column vector is being collected, the number of elements collected by **getdoubles()** can be limited to one. **getdoubles()** returns, as the value of the function, the number of **double**s collected.

These considerations lead to the C code for **pgetv()**:

```
/*
***********************
pgetv-get a vector from fin.
***********************
*/

int pgetv(fin,fout,v)

  FILE *fin,*fout;
  double v[];

{
  int finished = NO,len = 0;
  int n = 0,bound = MAXVEC,first_time = YES;
  int getdoubles();

/*
loop until a valid vector is read
*/

    while( finished != YES)
    {

/*
get a row of doubles from input
*/

      if( (n = getdoubles(fin,fout,v+len,bound)) == EOF)
        return(EOF);

/*
if first time, then check if this is a row vector
  and set the upper bound appropriately
*/

      else if( first_time == YES && n > 0)
      {
        first_time = NO;
        bound = (n > 1 ? 0:1);
        len = n;
      }
```

```
/*
if something already collected but nothing returned,
  then finished
*/

    else if( n == 0)
      finished = (len > 0 ? YES:NO);

/*
otherwise increment len and check if bound
  should be reset
*/

    else
    {
      len += n;
      bound = ( len < MAXVEC ? bound:0);
    }
  }

    return(len);
}
/*
end of pgetv
*/
```

Initially, **bound** is set to **MAXVEC**, since this is the maximum number of elements that a vector can have. Depending on how many elements were read on the first iteration, **bound** is set appropriately to limit the next input. If no elements were read on the first iteration, neither **first_time** nor **bound** should change. If the first call to **getdoubles()** read more than one element, a column vector is out of the question and the next iteration should read no elements (i.e., a blank line terminating the vector), so **bound** is set to zero. If the first call read only one element, then a column vector could be collected, and a maximum of one element can be read during the second iteration. In the latter case, **bound** is set to one. A conditional expression is used to compactly make this distinction and set **bound** in a single line. **finished** is also set by a conditional expression, if no elements were returned from **getdoubles()** and a vector is already collected. Conditional expressions were discussed in Section 2.14.

The starting address of the **double** buffer **v[]** is passed to **getdoubles()** using the pointer form of indexing. If **len** elements have already been collected, then **v + len** is a pointer to the first empty slot in **v**. The conditional expression checking if **len** is less than **MAXVEC** is needed to be sure that no more than **MAXVEC** elements are entered. If **v[]** has no more empty slots, then **bound** is set to zero so that **getdoubles()** won't accept any elements on the next iteration.

The code for **pgetm()** is similar to **pgetv()**, and is therefore not presented. About the only major difference is that **pgetm()** must check each row to be sure that the number of elements matches the number in the first row, which determines the row size for the entire matrix. In addition, the logic for dealing with column vectors is not needed, though the **bound** parameter to **getdoubles()** can be used to limit the maximum number of rows collected to **MAXVEC**, as well as to limit the size of each row.

3.5.1. Exercises

1.   Write **pgetm()** using **pgetv()** as a model.

3.6. getdoubles()

pgetv() and **pgetm()** are expecting **getdoubles()** to return a **double** buffer with zero or more **double**s, and the number of **double**s as the value of the function. The number of **double**s returned also must not exceed the value of the upper bound parameter passed to **getdoubles()**. **getdoubles()** and routines called by it must deal with converting character input to binary and the protocols for line extension and pipes, and must then report any errors.

A pseudo-code outline for **getdoubles** is:

```
while( doubles not yet collected )
{
    Get a buffer of doubles in character form from fin;

    if( end of file reached)
        return(EOF);

    while( there are prospective doubles )
    {
        Convert double from character to binary;

        if( no double converted)
        {
            Report error and start over;
        }

        else if( upper bound on number of elements exceeded)
        {
            Report error and start over;
        }
    }
}

return( number of elements converted);
```

To simplify the input task, **getdoubles**() only deals with converting a character buffer filled with prospective **double**s in character form into binary and lets another function, **getrow**(), take care of fetching the character buffer. As was the case for **pgetv**() in the last section, **getdoubles**() passes along the **FILE *** parameters to **getrow**(), along with a character buffer for **getrow**() to fill with prospective **double**s. The logic for dealing with the line extension and pipe protocol can therefore be isolated in **getrow**() and routines called by it.

Because the loop is relatively short, a **return** is made in the middle if **EOF** is read, rather than making **EOF** a special case for **finished** in the **while** condition. In a more complicated loop, setting a status variable would be more appropriate, so that the conditions for exiting the loop are isolated at the top or bottom, rather than being spread throughout the loop body.

The inner **while** loop also needs a way to separate out prospective **double**s from the character buffer and convert them into binary. The second task can be handled by the C library function **sscanf**(). The details of using **sscanf**() are discussed in Section 3.9, but for the moment all we need to know is that **sscanf**() operates similarly to **printf**(). The first parameter is a string, which is to be converted from character to binary, the second parameter is a format specification, similar to that of **printf**(), and the rest of the parameters are *pointers* to the variables into which the results of the character to binary conversion are to be put. Pointers are needed here because of C's call by value rule. If pointers are not passed, the results of the conversion won't be available to the calling routine.

The first part of the conversion procedure, separating out prospective **double**s from the character buffer, is accomplished using the function **getword**(). The job of **getword**() is to retrieve the next sequence of printable characters (a word) from the input buffer and deposit them in an output buffer. The three parameters for **getword**() are the position where **getword**() should look for the next word, the input buffer in which it should start looking, and the output buffer containing the result. The return value of **getword**() is the location in the input buffer where scanning for input stopped. Using **getword**(), the buffer filled with prospective **double**s can be sequentially broken into component words and the words can be converted into **double**s, if appropriate. Since **sscanf**() returns the number of items converted as the value of the function, we can check if a word was converted as a correct **double** if we check that **sscanf**() has indeed converted the character buffer to **double** form.

Here is the C code for **getdoubles**:

```
/*
**************************
getdoubles-get a row of doubles
    into the vector v, returning
    number of doubles fetched
    or EOF if end of file
    encountered.
**************************
*/

int getdoubles(fin,fout,v,bound)

  FILE *fin, *fout;
  int bound;
  double v[];

{
  char buf[MAXBUF],nxtwrd[MAXBUF];
  int idx,slen,sstrt,finished = NO;
  int getrow(),strlen(),getword();


/*
loop until a valid row of doubles converted
*/

    while(finished != YES)
    {

/*
get a row of doubles in character form
*/

      if( getrow(fin,fout,buf) == EOF)
        return(EOF);

/*
initialize idx, sstart, slen, and finished for the loop
*/

      idx = 0;
      slen = strlen(buf);
      sstrt = 0;
      finished = YES;
```

```
/*
convert doubles from character to binary
*/

     while( (sstrt = getword(sstrt,buf,nxtwrd)) < slen)
     {

/*
error if nondouble there
*/

        if( sscanf(nxtwrd,"%lf",&v[idx++]) != 1)
        {
          fprintf(stderr,"?invalid double.\n");
          fprintf(stderr,"%s\n",buf);
          finished = NO;
          break;
        }

/*
error if too many elements
*/

        else if( idx > bound)
        {
          fprintf(stderr,"?too many elements.\n");
          fprintf(stderr,"%s\n",buf);
          finished = NO;
          break;
        }
     }

  }

     return(idx);
}
/*
end of getdoubles
*/
```

In the C code for **getdoubles()**, several additional local variables are needed for controlling the call to **getword()** and access to the **double** buffer v. **sstrt** is the starting position in the input character buffer **buf**, and **slen** is the length of the input buffer. The inner loop continues until the index where **getword()** stopped scanning is beyond the end of string. The initializations for the inner loop take the optimistic viewpoint that conversion of the **double**s will succeed, hence **finished** is set to **YES**. The length of the input buffer, starting position for **getword()**, and the starting index in **v** are also initialized. The C library function **strlen()** returns the length of a null terminated string passed as a parameter.

On each iteration of the inner loop, the call to **sscanf()** tries to convert the word in **nxtwrd[]** into a **double**. If the conversion does not succeed, **sscanf()** returns zero, and an error message results. In the argument list to **sscanf()**, **idx**, the index of the next free slot in the **double** buffer **v**, is incremented using the autoincrement operator in the postfix position. This means that **idx** is used as an index before incrementing, so the proper slot in **v** is filled. In addition, the address operator, **&**, is used to get a pointer to the slot in **v** where the converted element should go. Another way of accomplishing the same thing would have been to use the pointer form of indexing for arrays, similar to the call to **getdoubles()** in the last section. If the conversion does succeed, **idx** is checked to be sure it has not exceeded **bound**. Either error causes the character buffer to be dumped with an error message and the loop returns to the call to **getrow()**, to get another character buffer.

3.7. getrow()

The task of **getrow()** is to read lines from **fin** and check whether they contain the vertical bar pipe symbol. If so, then everything after the vertical bar must be written to **fout**. **getrow()** leaves the dirty work of detecting whether an input line is continued with a backslash to **getbuf()**, which does the actual fetching of the input buffer, reading across input lines if a line is continued with a \:

```
/*
**********************************
getrow-get a row of matrix as characters
       from fin. Return EOF if end of
       file read. If PIPE_C appears,
       write rest of line to fout.
**********************************
*/

int getrow(fin,fout,buf)

  FILE *fin, *fout;
  char buf[];


{
  char *c;
  char *index();
  int getbuf();
```

```
/*
get a row from fin, continuing over end of
   line if necessary
*/

   if( getbuf(fin,buf) == EOF)
     return(EOF);

/*
if | appears, write out what's after it
*/

   if( (c = index(buf,PIPE_C)) != 0)
   {
     *c = 0;
     fprintf(fout,"%s",c+1);
   }

   return(OK);
}
/*
end of getrow
*/
```

The function **index()** is part of the C library. **index()** takes a character string and a character as parameters, and returns a character pointer to the first element in the buffer where a matching character occurs. If no matching character occurs, **index()** returns zero. On more recent UNIX systems, **index()** is called **strchr()**, but the parameters and return value are the same.

An **EOF** returned from **getbuf()** causes **getrow()** to return immediately. The vertical bar pipe symbol | is defined as the constant **PIPE_C**. **index()** is used to get a pointer to **PIPE_C**, if any, and everything after the pointer is written out using **fprintf()**. The pointer form of indexing is being used here, so that if **c** points to an array element containing **PIPE_C**, **c + 1** will point to the next element, which is where the text to be passed down the pipe starts.

3.8. getbuf()

At the base of the function-calling chain between **pgetv()** and the input file is **getbuf()**. **getbuf()** fills an input buffer, reading across newlines if the last character on the line is the backslash (\) character.

The pseudo-code for **getbuf()** contains a loop which reads in text lines until either the buffer is full or no backslash occurs as the last character before the newline in the input buffer:

Set **buf[0]** to \\ ;

Set the index, **n**, into **buf[]** to 0;

while(buf[n] is a backslash **&&** **buf[]** is not full)
{

    Read from **fin**, filling buffer starting at **buf[n]**;

    if(end of file was read)
        return(EOF);

    Reset **n** to the index of the last character before the newline;
}

return(OK);

The C code for **getbuf()** follows from the pseudo-code by condensing the two operations of reading the buffer and checking for end of file into one **if**:

```
/*
*************************************
getbuf-fill input buffer, reading across lines
        if \ appears at end of line. Return EOF
        if end of file read.
*************************************
*/

int getbuf(fin,buf)

  FILE *fin;
  char buf[];

{
  int n=0;
  int strlen();
  char *fgets();

    buf[n] = CONT_C;

    while( buf[n] == CONT_C &&
        n < (MAXBUF - ENDOFFSET)
        )
    {
      if( fgets(buf+n,MAXBUF-n,fin) == NULL)
        return(EOF);
```

```
        n = strlen(buf) - ENDOFFSET;
    }

    return(OK);
}
/*
end of getbuf
*/
```

We've used **strlen()** to calculate the length of the input buffer. The length is then used in the calculation of the index where the next input line should be started, in the event a backslash appeared at the end of the line. Since both the newline and the backslash must not appear in the final buffer, the first character in the next line read should be placed into the same array position as the backslash, causing further input to overwrite the backslash and newline characters. The backslash character is defined as the constant **CONT_C**. **n** is the index of the backslash while **buf+n** is a **char** pointer to the element of the input buffer where the backslash character is located. The constant **ENDOFFSET** is the distance between the end of the input buffer string and the backslash:

$$\textbf{buf[n]:}\quad\textbf{buf[n+1]:}\quad\textbf{buf[n+2]:}$$

$$\textbf{\textbackslash}\qquad\quad\textbf{\textbackslash n}\qquad\quad\textbf{\textbackslash 000}$$

Because of C's zero-based indexing system, the length of the string returned by **strlen()** is one more than the index to the last character. The backslash is the second last character, since **fgets()** returns strings with the newline still attached. **ENDOFFSET** should therefore be 2.

An alternative implementation would have been to use a **for** loop instead of a **while** since the initialization, continuation, and re-initialization statements are clearly available:

```
for( buf[n] = CONT_C;
     buf[n] == CONT_C && n < (MAXBUF - ENDOFFSET);
     n = strlen(buf) - ENDOFFSET
   )
```

The length of the continuation expression makes the **for** loop header sloppier, though the loop control code is all isolated at the top.

One potential input error not caught here involves truncation of an input string because so many characters were read that all wouldn't fit into **buf[]**. In this case, the calculation involving **ENDOFFSET** would be wrong. If, however, **MAXBUF** is set large enough for any reasonably sized input line, then the problem should occur seldom enough that extra code for a check isn't warranted. To be absolutely safe, an additional **if** statement could be inserted to calculate **ENDOFFSET**, depending on whether a newline is there or not.

3.9. sscanf()

The C library function **sscanf()** takes a character string with text to be converted into binary form, a format string similar to (but not exactly like) that of **printf()**, and pointers to **double**s, **float**s, **int**s, and **char**s which are to receive the final binary. The call looks like:

sscanf(<buffer>,<format>,<param1>,<param2>, ... , <paramn>);

In general, **sscanf()** converts blank separated fields, like the character representation for **double**s in a vector, from a character string into binary and puts the result into the parameters. The number and type of parameters must match the format specification exactly as in **printf()**. The format string uses the same conventions as **printf()** with the following exceptions:

1. Blanks, tabs, and newlines in the format string are simply ignored.

2. Conversion specifications in the format string match characters in the character string which are not blanks, tabs, or newlines. In general, blanks tabs and newlines serve as *field separators* in the character string, like they do in the strings scanned by **getword()**. The nonblank characters are converted into binary according to the format specification. An exception to this rule is in 5 below.

3. An optional assignment suppression character can be used to cause **sscanf()** to skip over a particular field in the character string. The assignment suppression character is * and appears between the % and the field width, as in **%*c**, which causes the next character to be skipped.

4. The field width is a maximum width rather than a minimum as in **printf()**. **sscanf()** will read until the field width is exhausted or until the next blank, tab, or newline character is encountered. In addition, there is no precision, as in **printf()**.

5. The **c** (character) conversion causes the next character to be read from the character string and put in a parameter (unless assignment is suppressed) regardless of whether it is a blank, tab, or newline. If blanks, tabs, and newlines should be skipped, **%1s**
    (for a single character string) should be used.

6. There is no **g** conversion specification. **double**s must be read using **%lf** or **%le** (for long **f** and long **e**, respectively). **float**s are read using **%f** or **%e**. The distinction between **float**s and **double**s must be made, though either the **f** or **e** conversion character can be used for each.

For example, the call:

```
int n;
double x[2];
char s[70];

sscanf(
    "555 32.5 44.1 The end",
    "%d %lf %lf %s",
    &n,x,x+1,s
);
```

would read **555** into **n**, **32.5** into **x[0]**, **44.1** into **x[1]**, and "The", terminated with a null byte, into **s**. The entire string "**The end**" is not converted because **sscanf()** only reads up to the next blank. Notice that a pointer to **n** is passed to the function by using the address operator, **&**. Since **x** is an array, and since the names of arrays are pointers to the first element, **x** and **x+1** can be used to pass pointers to **sscanf()** for the first and second elements of **x**.

If the format string had read:

"%*d %lf %lf %s"

and the **n** had been omitted, the field containing the **555** would have been skipped.

By far, the most common mistake beginners make when using **sscanf()** is forgetting to pass pointers as parameters when conversion of scalars is desired. **sscanf()** is a function, and like any other function in C, all scalars are passed by value only. If a scalar is passed as a parameter to **sscanf()**, the value assigned to the scalar will be local to **sscanf()** and won't be accessible to the calling function. **sscanf()** also returns the number of items successfully converted as the value of the function. This can be checked to see how many parameters were assigned correctly.

sscanf() has two relatives, **scanf()** and **fscanf()**, which take input directly from files rather than from strings. **scanf()** reads directly from **stdin** and requires no **FILE** pointer, while **fscanf()** reads from a file whose **FILE** pointer is passed as the first parameter. Both functions operate exactly like **sscanf()**. They are less useful, however, since both will read across line boundaries in order to fulfill the format string. If the program is doing line-oriented i/o, this is a real inconvenience, since neither function can guarantee that a line of input will be scanned intact. Using **fgets()** to fetch a string and **sscanf()** to decode it usually leads to more predictable results.

3.9.1. Exercises

1. Write a program which uses **scanf()** to read vectors from **stdin** and **printf()** to echo them to **stdout**.

2. Write a program called **maxrain** which reads rainfall records taken at a series of rain gauges from **stdin**, converts them into binary form using **sscanf()**, and only writes them out if the rainfall amount exceeded 10 inches. The rainfall records have the following form:

    mm/dd/yr <station number> <station name> <rainfall amount>

    The first field is the date. Skip the station number while reading the data in, and be sure to write the name of the station and date as well as the amount.

3.10. getword()

getword() provides a way of moving along the text string number by number, so that, each time **sscanf()** is called, it can be given the next number in the character string to convert. Since an arbitrary number of blanks and tabs can separate the numbers, we cannot just skip the next character after a number and continue. We need a function which will return a nonblank word from a character buffer, given the position in the buffer from which it should start searching.

One quick and dirty way of implementing **getword()** would be simply to make a call to **sscanf()**, transferring the next nonblank string into the buffer, then calculating the index in the source string where copying stopped, using the length of the nonblank word and the source string length. Complications arise when we try to account for blank spaces in front of the word, since **sscanf()** skips blanks without telling how many are there. Rather than try to deal with these complications, we'll implement **getword()** by having it straightforwardly copy characters from the source string into the destination.

getword() takes as parameters an index in a character string where it should start, a character string to be processed, and a buffer where the result should be placed. **getword()** copies the next nonblank word from the character string into the buffer, skipping any leading blanks, and returns the index of the character in the source string where it stopped copying. This can be expressed in pseudo-code as:

    while(the source string has a blank or tab)
    {
        Increment the index into the source string;
    }

    while(the current source char is not a blank or tab &&
           the source string is not at the end
)
    {

Copy a character from the source string into the destination;

Increment the indices into the destination buffer and the source string;
}

Null terminate the buffer;

return(index in the source where copying stopped);

Whether a character in the source string is a blank or tab can be determined by the C macro **isspace()**, which is defined in the system **#include** file **ctype.h**. **isspace()** evaluates to zero if the character it is passed as its single parameter is not a blank, tab, carriage return, newline, vertical tab, or form feed, and a nonzero result otherwise. **ctype.h** must be included in the file where **getword()** is defined, using the **#include** statement. As with the macros **getc()** and **getchar()** discussed in Chapter 2, **isspace()** should not be declared, since it is not a function.

The code for **getword()** follows straightforwardly from the pseudo-code specification:

```
/*
*******************************************
getword-get next nonblank word from buf1 to buf2,
         starting at i. Return the index in buf1
         after the word copied.
*******************************************
*/

int getword(i,buf1,buf2)

  int i;
  char buf1[],buf2[];

{
  int j=0;

    if( i < MAXBUF)
    {
      while( isspace(buf1[i]) &&
          buf1[i++] != 0
        )
      ;

      while( !isspace(buf1[i]) &&
          (buf2[j++] = buf1[i++]) != 0
        )
```

```
            ;
        }

    buf2[j++] = 0;

    return(i);
}
/*
end of getword
*/
```

The first **while** loop indexes over any leading blanks in the string, while the second copies characters into the buffer until the first whitespace character occurs. The body of both **while** loops is empty, since all the incrementing and copying is done within the **while** condition. The autoincrement operator in the postfix position causes the the index to be used before being incremented, so that the correct slot in **buf1[]** is tested in the first loop and the correct character is copied from **buf1[]** to **buf2[]** in the second. **j** is not checked to be sure it does not run over **MAXBUF** because **j** is bounded by **i**, and **i** is checked at the start of the function. A null byte is deposited into the end of **buf2[]** before the function returns, so that it constitutes a proper character string.

Because **getword()** is potentially useful in a wide variety of programs, it is a candidate for inclusion in **c.c.** As was discussed in Section 2.17, **c.c** contains functions which are supplemental to the C library, and have potential usefulness beyond a single program or group of programs. We'll have occasion to use **getword()** again.

3.10.1. Exercises

1.   Rewrite **getword()** and **getrow()** to use **sscanf()** directly.

3.11. The Onion Principle

Most good computer programs are like onions. The outer layers can be stripped away to reveal inner layers, which can be further stripped away, etc., until only library and system calls (like **printf()**) remain. The "onion principle" can be best illustrated by a partial call graph of **dotp**:

```
main()
>>getvecs()
>>>>pgetv()
>>>>>>getdoubles()
>>>>>>>>strlen()
>>>>>>>>getrow()
>>>>>>>>>>fprintf()
>>>>>>>>>>index()
>>>>>>>>>>getbuf()
```

```
>>>>>>>>>>>>fgets()
>>>>>>>>>>>>fprintf()
>>>>>>getword()
>>>>>>>>isspace()
>>vdotp()
>>printf()
>>exit()
```

A call graph is sometimes more useful than a flow chart, since it shows the hierarchical nature of a program without the clutter of implementation details. If each function has a clearly defined purpose, then a call graph is the equivalent of a functional flow chart. It suppresses the details of the program's code without hiding the overall flow of the program. Call graphs are especially helpful for C programs, since functions tend to be small. The call graphs for **addm** and **eqsolv** would show a similar calling structure below the routines **pgetv()** and **pgetm()**.

In general, the inner layers of the onion tend to be concerned more with particular implementation details than the outer layers. C library functions and system calls, like **fopen()**, are good examples. Depending on the nature of the hardware, the inner structure of **fopen()** might differ radically from one implementation of the UNIX system to another. The syntax for a particular command language or a data interface is another implementation detail which should be pushed as far down into the onion as possible. The details of the vector-matrix data exchange protocol are hidden beneath the functions **pgetv()** and **pgetm()**, on the input side, and **fputv()** and **fputm()**, on the output side. These routines could be completely rewritten, as long as they fulfilled the expectations built into their calling routines.

The advantage of hiding details in the inner layers is that drastic changes can be made in the implementation without perturbing the outer layers much at all. For example, the vector-matrix data interface described in this chapter underwent five major revisions before taking on the form presented here. Each revision involved only the routines from **pgetv()** and **pgetm()** on down. The upper-level routines and numerical functions changed very little. For small programs like **dotp** and **addm**, the isolation provided by code layering is a nice convenience, but the real payoff comes when you're working with a system containing thousands or hundreds of thousands of source lines. Unless the code is hierarchically structured, such systems swiftly become an unmanageable mess.

The disadvantage of modularization is that more function calls and run time execution steps are necessary to obtain results, potentially slowing down execution. From the call graph for **dotp**, it is obvious that a lot of code is involved in getting vectors from a file. This code would not be necessary if a straightforward, FORTRAN-like READ-FORMAT interface were used; however, the flexibility of the vector-matrix pipe protocol would be lost. For small programs like **dotp** and the other vector matrix tools, which do a limited amount of i/o, the increase in execution speed brought by abandoning the vector-matrix protocol is probably not worth the loss in flexibility, but for large applications which require transferring huge amounts of data quickly, a faster interface is probably more desirable.

3.12. References

Two of the best references on program structuring are:

1. *Software Tools*, by Brian W. Kernighan and P.J. Plauger, Addison-Wesley, Reading, MA, 352 pp., 1976.

2. *Software Tools in Pascal*, by Brian W. Kernighan and P.J. Plauger, Addison-Wesley, Reading, MA, 368 pp., 1981.

Kernighan and Plauger discuss layering of code, information hiding, and many other aspects of software engineering which are important for making code easy to implement and maintain. The language used in *Software Tools* is Ratfor, a superset of FORTRAN with a syntax much like that of C. Both books are primarily concerned with systems software, like editors and macroprocessors, but numerical programmers can benefit from reading them also.

Another source for information and tips on good software engineering practice is a column called "Programming Pearls" which appears periodically in the journal *Communications of the ACM*. Although the topics generally tend to stress system programming applications, the basic content is often more widely applicable.

The discussion in Section 3.3 on output precision for tools exchanging data through pipes touched on the need to use an output precision which is equal to or greater than the floating point precision of the machine. For further information about floating point precision, and about how to measure the characteristics of a machine's floating point unit, see:

1. *The Engineering of Numerical Software*, by Webb Miller, Prentice-Hall, Englewood Cliffs, NJ, 176 pp., 1984.

Although the programming examples are in FORTRAN, they can easily be translated into C.

4

A GRAPHICAL INTERLUDE

Plots are often very effective for communicating the results of a numerical study. A plot can summarize data in a way that columns of numbers in tabular form cannot. In addition, plots can suggest paths for further analysis, which might be overlooked if just the raw numbers were available. A simple tool for plotting two variables against each other can serve as the workhorse for much engineering and scientific graphics work. With such a tool, the scientist or engineer can interactively generate and analyze data at a terminal much more quickly than if batch plotting routines were used.

In this chapter, a simple graphics command language is interpreted by a graphics filter, **graphit**, into graphical operations on a peripheral device. The language includes basic operations to draw lines, move the graphics cursor or pen from one point to another, change color (if any), and map data points into screen windows. The command language is integrated with the vector-matrix protocol of Chapter 3, so that, with the addition of a starting and an ending command, a sequence of vectors can be plotted as lines.

Unfortunately, there is little uniformity in the way graphical output devices respond to commands. Many low-performance terminal type devices accept sequences of nonprinting ASCII characters which they translate into vectors. Some plotters have their very own command languages, similar to assembly language, that are translated into pen movements. High-performance memory mapped devices and bit mapped personal computer screens are often accessible by directly changing the bit values at a memory location corresponding to a pixel on the screen.

The differences are usually hidden behind a graphics subroutine library. The subroutine calls for various devices remain the same from device to device, though different libraries must be linked into the program for different devices. These libraries are often implemented in FORTRAN or Pascal on minicomputers and mainframes, or in assembly language with a

BASIC interface on microcomputers. Rarely are they implemented directly in C.

We'll assume that some generically named C functions can be written to interface with the graphics subroutine package. These functions are all prefixed with **gr**, like **grline()**. The majority of the **gr** functions are straightforward enough and dependent enough on the local graphics library that they would need to be designed with a particular graphics library in mind. Translating color names into color changes is an exception, however. To illustrate how the mapping from color names to colors is achieved, we'll assume that the library function call to change color takes an integer argument, ranging from 0 to 7, indicating the new color. This requires implementing some code for the color-changing function, **grcolor()**.

In addition, the last section will discuss foreign language function calling at the source level, so that foreign language graphics libraries can be linked with **graphit**. Of course, this presupposes that your C implementation and linker can handle foreign language calling, a facility which most multilanguage machines have. Foreign language calling at the source level is highly machine and implementation dependent, so only general pointers can be given. Note that the suggestions given in the last section can also be applied to using existing numerical subroutine library routines written in FORTRAN with the numerical tools.

The advantage of using **graphit** over writing calls to graphics routines directly into numerical programs is that the graphics programming only needs to be done once. Other programs can communicate to **graphit** through pipes if graphical output is desired. The user can experiment with different positions of plots on the screen, different colors, etc. without having to recode, recompile, and relink the numerical program. In contrast, if the graphical routines are wired directly into the code for the numerical program, the graphical output section would need to be rewritten for each new application.

4.1. The Graphical Language

The heart of **graphit** is a command language which is interpreted to call the action routine that performs a particular graphical task. **graphit** is invoked from the UNIX shell in the following manner:

$ **graphit**

graphit reads text lines from **stdin**, trying to interpret lines which begin with an ''at'' sign (**@**) as a command. If the line does not begin with **@**, it is written to the graphics device without change. Each command line consists of **@**, followed by the command. After the command comes a list of zero or more blank separated command parameters. An example is:

@color blue

This command would cause **graphit** to call the routine **grcolor()** with the argument string ''**blue**''. **grcolor()** would decode the string and call the device-dependent graphics subroutine library function with an appropriate argument.

An exception to this command format is the command to draw a line. We want **graphit** to accept vectors for the line endpoints in the format specified by the vector-matrix protocol from Chapter 3, but we can't simply have **graphit** read vectors by using **pgetv()** all the time, since commands would then be ignored. Remember, **pgetv()** always returns with a correct vector or **EOF**, flagging all other input as an error. We also need to control which co-ordinates of a vector get plotted against each other to form a line. If **graphit** reads a series of five element vectors from **stdin**, only two of the elements can be used as co-ordinates for line endpoints, since the display or paper is two dimensional.

A line drawing sequence starts with the **@line** command, which takes two parameters. The parameters are the indices of the elements in the vectors to follow which are to be used as endpoints for lines. The vectors come next, followed by a single vector with -1 as its only element. Since vectors which are to be used as line endpoints must have at least two elements, a vector with the single element -1 is easily distinguishable from valid input.

Here is an example of a series of commands, or script, which draws a square whose corners are at (0, 0), (1, 0), (1, 1), (0, 1):

```
@line 1 3
0 0 0 0

0 1 0 0

0 1 0 1

0 0 0 1

0 0 0 0

-1
```

This sequence uses the second element of the vectors as the x co-ordinate and the fourth element as the y co-ordinate of the square's vertices. An example of the output from this sequence is shown in Fig. 4.1. The **@to** command, which moves the graphics cursor or pen without drawing a line, has the same command format as **@line**.

graphit uses a mathematical transformation to take points in a user-defined data window and transform them into a user-defined viewport on the plotting surface. This procedure is often called a window-viewport transformation. Incoming data points and lines are clipped to a window in data co-ordinates. Anything outside the window is not displayed. The data points are then mathematically transformed into plotting surface co-ordinates inside a plotting surface viewport. The plotting surface viewport may cover all or part of the physical plotting surface. Fig. 4.3 in Section 4.6 illustrates this process in more detail. The data window limits what data gets plotted on the screen, since the all data to be plotted are first clipped to this window. The plotting surface viewport determines where on the screen the plot is located, since the points which are inside the data window are projected onto the plotting surface within the viewport. Clipping assures that lines which project outside the data window are not displayed. The clipping and window-viewport algorithms are discussed in Sections 4.5 and 4.6,

Figure 4.1. Output from the Square Drawing Script

respectively.

In order to keep the presentation simple, the command language for **graphit** is limited to a few basic operations of widespread applicability. More complex commands can be built by combining these commands, or by introducing others to take advantage of special features available with certain peripherals. **graphit** understands ten commands:

1. **@view** $x_1\ y_1\ x_2\ y_2$

    Set the plotting surface viewport so that the lower left corner of the viewport is at (x_1, y_1), while the upper right corner of the viewport is at (x_2, y_2). The point co-ordinates are reals with $x_1 \leq x_2$ and $y_1 \leq y_2$. Initially, the viewport covers the entire plotting surface. This command is used in conjunction with the **@wind** command.

2. **@wind** $x_1\ y_1\ x_2\ y_2$

    Set the data window so that the lower left corner is at (x_1, y_1), and the upper right corner is at (x_2, y_2). The point co-ordinates are reals with $x_1 \leq x_2$ and $y_1 \leq y_2$. Initially, corners of the data window are set to large negative and positive values, to simulate infinity. Incoming points are clipped to the window boundaries using a clipping algorithm and transformed into viewport co-ordinates using a window-viewport transformation. The window-viewport transformation and clipping algorithms are discussed in later sections.

3.   **@line** <x index> <y index>

This command specifies that the elements <x index> and <y index> of the vectors to follow should be used as endpoints for pen movement, with the pen down. The sequence of vectors to be used follows, and the connected line or pen movement is ended with a vector whose single element is -1.

4.   **@to** <x index> <y index>

Same as for the **@line** command, except raise the pen first, so no line is drawn.

5.   **@color** <color name>

Change the color used to draw lines to <color name>.

6.   **@clear**

Clear the screen of graphics and alphanumeric characters. For a plotter, this command has the effect of advancing the paper to the next page, and reinitializing for a new plot.

7.   **@newline**

Move the alphanumeric cursor to the far left column of the display or page. This command is especially helpful for interactively positioning alphanumeric text, since newline characters are not automatically written out at the end of a line of text. For plotters, this command may be undefined.

8.   **@graphic**

Switch the terminal or plotter into line drawing mode.

9.   **@alpha**

Switch the terminal or plotter into alphanumeric mode.

10. **@file** <filename>

Take further input from <filename> instead of from **stdin**. This lets you prepare figures (like plot axes or icons) in files beforehand and combine them into a single plot.

4.1.1. Exercises

1. What additional numerical tools would you need to be able to automatically draw and label a pair of axes using **graphit**? Should this capability be built into **graphit**? If you think so, develop a command syntax which will do it.

2. What other commands could be added to **graphit** to take advantage of hardware characteristics your graphic peripheral might have?

4.2. The Main Routine for graphit

The main routine for **graphit** is a simple command interpretation loop. A line of input is read from **stdin** into an input buffer. If the input buffer contains a command, the appropriate action for that command is taken. If not, the input buffer is written without change. This loop is continued until the end of file is encountered:

```
while( not at end of file)
{
      Read next line into input buffer;

      if( first character in the input buffer is not @)
      {
            Write input buffer;
      }
      else if( the next word in the input buffer matches a command)
      {
            Do the action appropriate for that command;
      }
      else
      {
            Write the input buffer;
      }
}
```

The inner part of the loop is actually a multiple **if-else** statement, in which a different function is called for each command. The pseudo-code will work fine as long as input is coming from **stdin**. But if an **@file** command happens to switch files, the program will exit as soon as the file ends instead of going back to reading from the previous file. We could modify the loop so that it switches back to **stdin**, but that still won't solve the problem if a nested **@file** command is encountered while reading from another file. What we need is a way to switch back to the immediately preceding file.

In order to deal correctly with nested **@file** commands, we'll put the command loop into a function called **interpret()** which has a single parameter: the **FILE *** from which to take input. The **main()** then becomes a four-line function which initializes the graphics device, calls interpret() with **stdin** as the parameter, closes the graphics device when **interpret()** returns, and finally exits.

```
/*
*******************************************
main-initialize graphics, start command interpreter,
     and close graphics when done.
*******************************************
*/

main()

{

  grinit();

  interpret(stdin);

  grclose();

  exit(0);
}
/*
end of main
*/
```

Any device-specific initializations should be done in **grinit()**, and any clean up, closing devices, etc., should be done in **grclose()**.

When **interpret()** encounters an **@file** command, the file is opened (if possible), and **interpret()** calls itself recursively with the new **FILE *** as the parameter. When input is finished, the call to **interpret()** returns to the previous context and input continues to come from the old file again. If the end of file is encountered while reading from **stdin**, **interpret()** returns to **main()** and the program exits.

In the following code for **interpret()**, a number of the **if-else** branches have been omitted to simplify presentation:

```
/*
*********************************
interpret-command interpretation loop.
*********************************
*/

interpret(fin)

  FILE *fin;

{
  FILE *fnew,*fopen();
  char buf[MAXBUF],file[MAXBUF];
  char *sgets();
  int n;
  int getcodes(),strncmp(),
      doto(),doline();

/*
loop, getting input until end of file
*/

    while(sgets(buf,MAXBUF,fin) != NULL)
    {

/*
not a command, so print it
*/

    if(buf[0] != COMCHAR)
      grprin(buf);

/*
interpret as command
*/


/*
@view command
*/

    else if(strncmp(buf+1,VIEW,NVIEW) == 0)
      doview(buf + NVIEW + 1);
```

```
/*
@line command
*/

    else if(strncmp(buf+1,LINE,NLINE) == 0)
    {
      if(doline(fin,buf + NVIEW + 1) == EOF)
        return;

    }

/*
@color command
*/

    else if(strncmp(buf+1,COLOR,NCOLOR) == 0)
      grcolor(buf + NCOLOR + 1);

/*
@file command
*/

    else if(strncmp(buf+1,CFILE,NCFILE) == 0)
    {

/*
get the name of the file and try to open
*/

        getword(0,buf + NCFILE + 1,file);

        if( (fnew = fopen(file,"r")) == NULL)
          fprintf(stderr,"graphit:can't open %s\n",file);

/*
recursively call interpret if opened ok
*/

    else
    {
      interpret(fnew);
      fclose(fnew);
    }
  }
```

```
/*
@clear command
*/

    else if(strncmp(buf+1,CLEAR,NCLEAR) == 0)
      grclear();


  . . .

/*
not recognized, so print buffer
*/

    else
      grprin(buf);

  }

  return;
}
/*
end of interpret
*/
```

The function **sgets()** simply calls the function **fgets()** and strips off the newline character at the end of the line. The code for **sgets()** is not presented here, because it simply involves a call to **fgets()**, and some minor further processing to remove the newline and insert a null byte. **sgets()** can be put into **c.c**, as was the case with **getword()** in Chapter 3, since it is useful in many programs.

The conditionals in the **if** statements use the C library function **strncmp()** to compare the input buffer to a string constant giving the command name. **strncmp()** is called with three parameters, the two strings to be compared (**string1** and **string2** in the following example) and the number of characters which are to be checked in the comparison (**n** in the example):

```
    int i,n;
    char string1[MAXBUF], string2[MAXBUF];

      i = strncmp(string1,string2,n);
```

strncmp() returns zero as the value of the function if the first **n** characters in both strings are equal. If the comparison is started after the **@** sign, **strncmp()** only compares the command string and ignores any parameters which are in the buffer after the command. In the command loop of **graphit**, the string length parameter **n** can be set to the size of the command string. **strncmp()** also has a relative, **strcmp()** which compares two strings completely, without any limit on the number of characters in the comparison. The calling statement for **strcmp()** is exactly the same as that for **strncmp()**, except that last argument (the number of characters) is omitted.

For the commands which take one or more parameters, like **@color** and **@line**, the implementing functions must decode the parameter as each command requires. **doline()** and **doto()** will return **EOF** if the end of file was encountered while reading vectors for plotting. If this occurs, **interpret()** immediately returns to processing the old file, or if called from **main()**, to the top level.

The implementing functions for **@newline**, **@alpha**, **@graphic**, and **@to** are **grnwlne()**, **gralpha()**, **grgraphic()**, and **doto()**. Calls to **grnwlne()**, **gralpha()**, and **grgraphic()** look exactly like the call to **grclear()**, while the calling statement for **doto()** looks like that for **doline()**.

The calculations used to identify the command name string and to skip it when one of the action functions are called require a bit of explanation. When the program enters the **if-else** portion of the loop, the first character in the input buffer is ***buf** or **buf[0]**. This character has already been identified as **COMCHAR** and must therefore be ignored in the string comparisons that follow. A pointer to the character in **buf[1]** is passed to **strncmp()** so that the initial **@** is skipped during the comparison. Likewise for the call to the routines implementing the command actions, both the **@** and the command name must be skipped. In order to do this, the size of the command name string (plus one for the **@** sign) must be added to **buf**, so that the pointer passed to the command routines points past the command name and the **@**.

In the code for the **@file** command, the file name parameter is removed from the end of **buf** and transferred into **file** by **getword()**. An attempt is then made to open the file in read only mode. If the file cannot be opened, an error message is printed to **stderr**, and processing continues with the old input file. If the file is opened successfully, a recursive call to **interpret()** occurs with the new **FILE *** as the parameter.

Each of the command name strings, their lengths, and the character identifying a command line are defined as constants. The constants are named to reflect their function. **ALPHA**, for example, is the command for changing the device from graphic to alphanumeric mode. The lengths of the command keywords are similarly named, as, for example, **NALPHA**. The **#include** file **graphit.h** contains preprocessor **#define** statements for all the command name strings, their lengths, and for **COMCHAR**:

```
/*
************ graphit.h **************
graphit.h contains constant declarations
  for graphit.
************************************
*/

/*
this character starts a command line
*/

#define COMCHAR     '@'

/*
graphics commands
*/

#define LINE        "line"
#define TO          "to"
#define COLOR       "color"
#define ALPHA       "alpha"
#define GRAPHIC     "graphic"
#define VIEW        "view"
#define WIND        "wind"
#define NEWLINE     "newline"
#define CLEAR       "clear"
#define CFILE       "file"


#define NLINE        (sizeof(LINE) - 1)
#define NTO          (sizeof(TO) - 1)
#define NCOLOR       (sizeof(COLOR) - 1)
#define NALPHA       (sizeof(ALPHA) - 1)
#define NGRAPHIC     (sizeof(GRAPHIC) - 1)
#define NVIEW        (sizeof(VIEW) - 1)
#define NWIND        (sizeof(WIND) - 1)
#define NNEWLINE     (sizeof(NEWLINE) - 1)
#define NCLEAR       (sizeof(CLEAR) - 1)
#define NCFILE       (sizeof(CFILE) - 1)
```

Other constants and macros specific to **graphit** can also be put into **graphit.h**. The keyword lengths are defined using the C **sizeof** operator, discussed in the next section.

The defined constants for the command keywords and their lengths make **interpret**() completely independent of the choice of command keyword names. If shorter keywords are desired, they could be substituted for the command definitions in **graphit.h** without having to touch the code in **main**(). Similarly, the **@** could be changed to another character.

4.2.1. Exercises

1. If your graphics device has more capabilities than are covered by **graphit**, install code in **interpret**() for interpreting commands which make use of these capabilities. Write the appropriate interface routines to call the graphics library routines which implement the capabilities.

4.3. sizeof

Since all the template command strings will be defined at compile time in **graphit.h**, savings in execution time can be realized if the lengths of the strings can be calculated at compile time too. The **sizeof** operator provides this facility. The C operator **sizeof** yields the size of its operand in bytes.

If the operand is a string constant, as in this case, then the result of applying **sizeof** to the constant name is the number of characters in the string, plus one for the null byte at the end of the string. The lengths of the command strings are therefore calculated by subtracting one from the result of applying **sizeof** to the string. Applying **sizeof** to variables of the predefined types, like **int**, **double**, and **char**, yields the number of bytes the variable occupies. For a general array, applying **sizeof** to the array name yields the number of elements in the array times the size of each element. If the operand is a **struct** or has a type which was defined with **typedef** (like **matrix**), then the **sizeof** operator yields the amount of memory allocated to the operand. Since the **sizeof** operator is evaluated at compile time, it can be used without slowing the program down for a function call.

sizeof won't work if applied to a string or array which is dynamically built during execution of the program, with the intention of finding out how many elements were dynamically added. If the character array **s[]** is dimensioned for **MAXNUM** characters, then **sizeof** will return **MAXNUM**, regardless of how many characters are in the null terminated string, because **sizeof** counts the amount of space allocated at compile time, not the run time string size. Similarly, if **s[]** is passed to a function as a parameter, taking **sizeof(s)** in the function only gives the size of a character pointer, not the size of the allocated array, since only pointers are passed as formal parameters to functions if the actual parameter is an array name.

The operand for **sizeof** can also be a keyword for a predefined type or a programmer-defined type, like the name of a **struct** or a type name created with **typedef**, as, for example, **sizeof(double)** or **sizeof(matrix)**. If applied to a type, **sizeof** returns the size of a single element of that particular type, in bytes. For any variable, this is the number of bytes the compiler allocates to the variable when a declaration is made.

4.3.1. Exercises

1. Write a program which uses **sizeof** to calculate the size of **int**, **float**, **double**, and **char**, and print out the results.

2. Modify the program to print out the size of the **struct** used to define the **matrix** type and the **matrix** type itself.

3. If you are running on a microcomputer with limited memory, you might not want to allocate **MAXVEC** x **MAXVEC** array elements at compile time for a variable of the **matrix** type, when the actual size of the matrices used as data is considerably less. An alternative is to dynamically allocate this memory when the matrix is read in, using **sizeof** and the C library function **malloc()**

    malloc() is called with a single argument, the size of the memory chunk you need, in bytes. It returns a pointer to the first element in the chunk, or **NULL** if no more memory is left to be allocated. Using **sizeof**, we can get a pointer to a **double** array with exactly the right size:

    ```
    int i, j;
    double *mat;
    char *malloc();

        mat = (double *)malloc( i * j * sizeof(double));
    ```

    The pointer to be allocated is declared as a pointer to **double**s, since it will be a block of storage in which the **double**s are stored. If **i** and **j** are the row and column dimensions of the matrix, then this code fragment will return a pointer to enough memory to store the matrix. The type of the pointer returned by **malloc()** is a pointer to a **char**, so the type must be coerced into the proper type for the variable **m**. In C, a device called a cast is used to do this coercion. It is achieved by putting the type into which the expression is to be coerced in parentheses before the expression.

    Redefining the **matrix** type to use a **double** pointer:

    ```
    typedef struct
    {
      int row,col;
      double *mat;
    } matrix;
    ```

    the size of storage allocated will depend on the actual size of the matrix.

    We cannot use normal array indexing to access elements in the matrix, because the column dimension is no longer available at compile time. Instead, we can define a macro which essentially performs the indexing calculation on the **double** vector:

#define MREF(m, i, j)   (*((m)->mat + (i) + ((j) * (m)->col)))

All three arguments to the macro are enclosed in parentheses, since any one could be an expression which would need to be evaluated before being used. Note that this macro requires that **m** be a pointer to a **matrix**.

This macro can be used on either the left or right side of an assignment expression:

> **matrix *m;**
> **double x,y;**
>
> **x = MREF(m,2,7);**
> **MREF(m,5,4) = y;**

After the program no longer has any use for storage allocated using **malloc()**, it should dispose of that storage by calling **free()**:

> **free(m->mat);**

Note that **free()** should only be called with a pointer to storage allocated using **malloc()** as an argument; otherwise, confusion and chaos could ensue.

Rewrite the vector-matrix i/o package from Chapter 3 using dynamic storage allocation.

4.   In addition to allocating storage for the two-dimensional array, it is also possible to dynamically allocate storage for the **matrix struct** itself. Add dynamic allocation for the **matrix struct** to the vector-matrix i/o package.

4.4. Storing Window-Viewport and Color Table Information

The routines **doview()** and **dowind()** are responsible for converting the viewport and window corners to binary form and storing them somewhere until they are needed in **doline()** and **doto()** for clipping and co-ordinate transformation. Since **doview()**, **dowind()**, **doline()**, and **doto()** are separate routines, global variables are needed to communicate this information between them.

One implementation strategy would be to store the information in the kind of program-wide globals which were discussed in Chapter 1. However, only the window-viewport changing routines and the plotting routines require information on the window-viewport corner points. It makes more sense to confine the visibility of this data to the routines that actually need it. Such modularization helps reduce data interconnections to a minimum, and can often cut down on mysterious bugs resulting from a global and a local with the same name.

Restricting the visibility of variables increases readability too, since a reader doesn't always have to refer back to a list of globals for the entire program. The overall result is that program maintenance becomes much simpler.

C provides some excellent facilities for controlling the extent or scope of a variable's visibility. In addition to the "visible everywhere" type global discussed in Chapter 1, C has a global storage class called **static**. A **static** global is visible only to routines defined in the same file where it is declared. Outside that file, it cannot be accessed. Since the scope of the variable's visibility is restricted to the file in which it is defined, **static** globals are said to have file scope. The **struct** template definitions and **typedef** statements discussed in Section 2.10 also have file scope. Program-wide globals and **static** globals are also called external variables, since they are external to any function. The **extern** keyword should therefore be used in the variable declarations at the top of functions in which global and **static** external variables are accessed, as was discussed for program-wide globals in Chapter 1.

Letting the corner points for the viewport be **vll[]** (lower left) and **vur[]** (upper right), and for the window be **wll[]** and **wur[]**, the **static** storage class can be used to restrict visibility to the file where **doview()** and the other routines using the window-viewport data are defined:

```
/*
viewport corners
*/

static double vll[DIM], vur[DIM];

/*
window corners
*/

static double wll[DIM], wur[DIM];
```

static variables imply a sense of privacy, since routines outside the file cannot access them. Functions in the file which need to access these variables should declare them **extern** in the variable declaration section:

```
int dowind(buf)

char *buf;

{
extern double vll[],vur[];
extern double wll[],wur[];

. . .

}
```

The **static** storage class can also be used for local variables within a function. The difference between normal local variables (often called ''automatic'' variables) and **static**s is that **static**s hang around after the function call exits and are guaranteed to have the last value they were assigned when the function is entered again. Local **static**s can be used to save information from one function call to the next. The values of automatic variables, on the other hand, are formally undefined (and are probably garbage) until they are specifically initialized.

A local **static** is declared just like a local automatic, except the type name is preceded by the **static** keyword:

```
int f1(a,b,c)

int a[],b;
char c[];

{
  static char buf[MAXBUF];

  . . .

}
```

Local **static**s can be used in our version of **grcolor()** to provide a lookup table for matching the color names in command parameter character strings with the integer control codes for changing the colors. Since the data require that we store a group of character strings corresponding to the color names, we'll use an array of pointers to the strings:

```
int grcolor(buf)

char buf[];

{
  static char *ctable[MAXTABLE];

  . . .

}
```

Pointers to strings are used here instead of a two-dimensional array like:

```
static char ctable[20][20];
```

because the names of the strings will have different lengths. It makes no sense to allocate space for a 20-by-20 array of characters when only one out of the 20 strings has a full 20 characters. An array of pointers to strings allows you to build a ''ragged array,'' in which the rows have different lengths. Each array element points to a character string with the name of a particular color. The array index is the same as the integer color code which is passed to the graphics library routine for changing colors.

Unlike automatic local arrays, **static** local, **static** global, and global arrays can be initialized at compile time. Arrays are initialized in C by listing the components after the array declaration, separated by commas, and enclosing the entire list in curly brackets with an = between the array name and the opening bracket:

```
static char *ctable[] =
{
  "black",
  "blue",
  "green",
  "yellow",
  "orange",
  "magenta",
  "red",
  "white",
  0
};
```

After the initialization, **ctable[]** contains pointers to eight character strings, with the last position in the array being a null pointer. Notice that the size of the pointer array wasn't specified. For an array which is initialized at compile time, the compiler will figure out how much space to use if you don't specify it. The null pointer is therefore needed in the last array slot to distinguish where the pointer array ends, and is a bit like the null byte used to terminate character strings. It can be checked in the conditional part of a **while** loop to determine when the contents of the pointer array have run out.

Global and **static** global arrays can be initialized using the same syntax:

```
static double vll[DIM] = { V_MINX, V_MINY};
static double vur[DIM] = { V_MAXX, V_MAXY};
```

The constants **V_MINX**, **V_MINY**, **V_MAXX**, and **V_MAXY** are defined using the **#define** preprocessor statement, so they are translated into numbers before the compilation begins. As long as the operands are defined at compile time, arithmetic operations can also be used in initialization statements. For example, the **vur[]** corner could have been defined using the minimum corner and an offset:

```
static double vll[DIM] =
{
  V_MINX,
  V_MINY
};
```

```
static double vur[DIM] =
{
  V_MINX + V_XOFF,
  V_MINY + V_YOFF
};
```

Arithmetic operations can be included in initialization statements for any variables, as long as all the operands are known at compile time. Of course, **static** scalars can be initialized too, just like automatic scalars.

Unlike automatic locals, however, **static** locals are initialized only once. As discussed in Section 2.3, automatic locals are initialized every time a function is entered. Because they are not permanent and are made anew each time a function is entered, automatic locals are not initialized at compile time. **static** locals, on the other hand, are only initialized at compile time, and will maintain their value between invocations of a function.

If the variable being initialized is a more complicated aggregate, like a **struct**, a similar syntax can be used. We may want to redefine our color lookup table to use a **struct** that maintains the association between the name and the integer code:

```
struct ctdef
{
  char *name;
  int code;
};
```

Defined constants like **BLACK, BLUE**, etc., can be used to represent the codes. An array of **struct**s can be defined and initialized to contain the name-integer code information:

```
static struct ctdef ctable[] =
{
  {"black",BLACK},
  {"blue",BLUE},
  {"green",GREEN},
  {"yellow",YELLOW},
  {"orange",ORANGE},
  {"magenta",MAGENTA},
  {"red",RED},
  {"white",WHITE}
};
```

In general, data objects which are aggregates of other objects, like arrays and **struct**s, are initialized by enclosing the inner objects in braces. The inner braces could actually have been omitted here, however, since the members of each **struct** in the array are integers and character strings, but they would have been required if one of the **struct** members had been another **struct**.

Nothing in the initialization statement requires the color constants (**BLACK, BLUE**, etc.) to be integers. If your device uses some other data to change colors, like a sequence of nonprinting or printing ASCII characters, preceded by an escape character, then the definition for **ctdef** and the **#define** statements for the color constants could be easily modified to accommodate the difference.

The compiler calculates how much space to allocate for the **struct** array, exactly as with the array of character pointers. Because the members of the array are **struct**s and not pointers, as in the character pointer array, we cannot set the last element to zero to indicate where the array stops, because zero isn't a valid data value for a **struct** array element. We could simply count the number of **struct**s in the array and use a defined constant to tell us how long the array is, but that would defeat the purpose of not dimensioning the array in the first place. If we recompiled **graphit** for another device with a wider range of colors, we would have to remember to change the **#define** statement for the defined constant.

What we need is a way to have the compiler calculate the number of elements in the array. As with the command strings, using the **sizeof** operator to define a constant provides that capability:

> **#define NCOLORS   (sizeof(ctable) / sizeof(struct ctdef))**

sizeof(ctable) calculates the size of the entire array. When the array size is divided by the size of a single element, resulting from applying the **sizeof** operator to the array type, the number of elements in the array results. In this way, the array size constant is dependent only on the declaration of the **struct** and the **struct** array and not on whether we remember to count the number of elements in the array when the number of colors changes.

4.5. The Clipping Algorithm

Clipping can be done either before or after a line or point has been transformed into viewpoint co-ordinates, though it usually makes more computational sense to clip before. If lots of lines are being clipped, then clipping them to the data window boundaries eliminates those lines which are completely invisible first, before the window-viewport transformation is applied. Much unnecessary computation is saved, which is why **doline**() and **doto**() clip first.

The clipping algorithm starts by determining whether the line or point lies entirely within the data window, and, if not, whether it can be rejected as lying completely outside. If neither of these tests is satisfied, the line is divided into two parts, and the tests are applied to each part. Since the window will always be a closed rectangle, a line is either completely inside the window, completely outside the window, or can be divided so that one segment can be rejected as being completely outside.

To facilitate the rejection test, the space outside the window is divided into eight quadrants, and each quadrant is given a four-bit binary code, as Fig. 4.2 shows. Each endpoint of the line to be clipped is assigned the four-bit code, corresponding to the region it is in. If the point is to the left of the window, the first bit is set; if it is to the right, the second bit is set; if it is below, the third bit is set; and if it is above, the fourth bit is set. If the four-bit codes for both endpoints are zero, then the line lies on the screen, and can be accepted.

Taking the logical intersection of the four-bit codes for both endpoints indicates whether the line is completely outside the window. The logical intersection, or logical "and," is performed by comparing the bits in both codes. If a particular bit is one in both codes, then that bit is one in the logical "and" of the two codes. If one or the other or both bits are zero, however, that bit is zero in the result. When the logical "and" of both four-bit codes is nonzero, then the line lies outside the window boundaries, and it can be rejected outright.

1001	1000	1010
0001	0000	0010
0101	1000	0110

Figure 4.2. Code Regions for the Line Endpoints

If neither of these tests applies, then the line must be subdivided. The simplest way of subdividing a line is to find the intersection with one side of the window and throw out the part that lies outside the window. The other part can be re-examined and clipping continued until both endpoints have four-bit codes which are zero.

The clipping algorithm pseudo-code is somewhat longer than usual, since there are a number of cases to consider:

Determine the four-bit code for the left endpoint;

Determine the four-bit code for the right endpoint;

while(one or the other endpoint is not in the window)
{
    if(the logical "and" of the two codes is not zero)
        return(NO);

    Pick a point which is not in the window for finding the intersection with the sides;

```
if( the point is to the left)
{
    The new x co-ordinate is the lower left corner x co-ordinate;

    Calculate the new y co-ordinate from the intersection of the left side with
    the line;
}
else if( the point is to the right)
{
    The new x co-ordinate is the upper right corner x co-ordinate;

    Calculate the new y co-ordinate from the intersection of the right side
    with the line;
}
else if( the point is below)
{
    The new y co-ordinate is the lower left corner y co-ordinate;

    Calculate the new x co-ordinate from the intersection of the lower side
    with the line;
}
else if( the point is above)
{
    The new y co-ordinate is the upper right corner y co-ordinate;

    Calculate the new x co-ordinate from the intersection of the upper side
    with the line;
}

if(the selected point was the left endpoint)
{
    Copy the new point into the left endpoint;

    Recalculate the left endpoint code;
}
else
{
    Copy the new point into the right endpoint;

    Recalculate the right endpoint code;
}
}

return(YES);
```

4.6. The Window-Viewport Algorithm

The window-viewport algorithm is a classic case in co-ordinate transformation. As Fig. 4.3 illustrates, the idea is to take the data points within a data window and transform them so that they fit within a screen viewport.

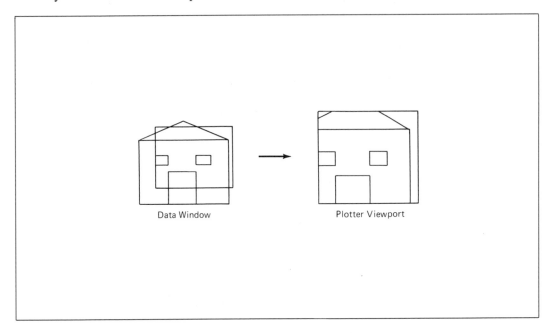

Figure 4.3. The Window-Viewport and Clipping Transformation

The lower left corner of the data window has co-ordinates (W_{lx}, W_{ly}) and the upper right corner has co-ordinates (W_{ux}, W_{uy}). The corresponding co-ordinates for the screen viewport are (V_{lx}, V_{ly}) for the lower left corner and (V_{ux}, V_{uy}) for the upper right corner. If (x_w, y_w) is a co-ordinate in the data window, then the transformation yielding the co-ordinates (x_v, y_v) in the screen viewport is:

$$x_v = (x_w - W_{lx})\frac{(V_{ux}-V_{lx})}{(W_{ux}-W_{lx})} + V_{lx}$$

$$y_v = (y_w - W_{ly})\frac{(V_{uy}-V_{ly})}{(W_{uy}-W_{ly})} + V_{ly}$$

In effect, the transformation is taking the distance between the window x or y co-ordinate and the lower left hand corner x or y co-ordinate, scaling this distance by the ratio of the viewport side to the window side, and adding the scaled distance (now in viewport co-ordinates) to the x or y co-ordinate of the viewport's lower left corner.

If we designate the scaling factors for the x and y co-ordinates by S_x and S_y, the equations can be rewritten:

$$x_v = (x_w - W_{lx})S_x + V_{lx}$$
$$y_v = (y_w - W_{ly})S_y + V_{ly}$$

Since all the data necessary to calculate S_x and S_y are known at the time the **@view** or **@wind** command is given, the scaling factors can be precalculated, and stored in another **static**. This avoids having to do the calculation every time a point is transformed, cutting down on the amount of execution time necessary to plot a line.

4.7. dowind() and doview()

dowind() and **doview()** are responsible for translating a user request for a change in the window or viewport corners into the internal data structures which maintain this information. **doview()** must also check whether the user supplied values are inside the physical plotting surface and issue an error message if they are not. Both routines check their input to be sure the lower left corner is indeed at the lower left, and not above or on the other side of the upper right corner.

The code for **dowind()**:

```
/*
viewport corners.
*/

static double vll[DIM] = {V_MINX,V_MINY},
            vur[DIM] = {V_MAXX,V_MAXY};

/*
window corners.
*/

static double wll[DIM]={SMALL,SMALL},
            wur[DIM]={BIG,BIG};

/*
scale factor
*/

static double sfact[DIM]=
{
  S_CALC(V_MINX,V_MAXX,SMALL,BIG),
  S_CALC(V_MINY,V_MAXY,SMALL,BIG)
};
```

```
/*
**********************************
dowind-reset the data window lower left
      and upper right corners.
**********************************
*/

int dowind(buf)

  char *buf;

{
  extern double vll[],vur[];
  extern double wll[],wur[];
  extern double sfact[];
  int sscanf();
  double lleft[DIM],uright[DIM];

/*
translate corner points to binary
*/

   if( sscanf(buf,
          "%lf %lf %lf %lf",
          &lleft[0],&lleft[1],
          &uright[0],&uright[1]
          ) != 4
     )
   {
     fprintf(stderr,
          "graphit:wrong @wind parameter format:%s.\n",
          buf
          );
     return(ERR);
   }

/*
check to be sure corners located correctly
*/
```

```
                  if(lleft[0] > uright[0] || lleft[1] > uright[1])
                    {
                      fprintf(stderr,
                             "graphit:corner points wrong for @wind:%s.\n",
                             buf
                             );
                      return(ERR);
                    }

         /*
         copy into static data structures
         */

                  vcopy(DIM,lleft,wll);
                  vcopy(DIM,uright,wur);

         /*
         recalculate the scale factors
         */

                  sfact[0] = S_CALC(vll[0],vur[0],wll[0],wur[0]);
                  sfact[1] = S_CALC(vll[1],vur[1],wll[1],wur[1]);

                  return(OK);
         }
         /*
         end of dowind
         */
```

The corner point parameters are stored in temporary vectors, until they can be checked for the proper location, and only copied into the **static**s when their correctness is assured. The window-viewport arrays are dimensioned using the preprocessor-defined constant **DIM**. The address operator **&** is used to obtain the array addresses for **sscanf()**. By C's rules of pointer arithmetic, the addresses could also have been calculated by adding the array index to the first element pointer represented by the array name. The function **vcopy()** simply copies one **double** vector into another. Although it's not in the C library, **vcopy()** is so simple that its code won't be specified.

The external **static sfact[]** contains the scaling factors S_x and S_y from the previous section. These must be recalculated every time the window or viewpoint corner points are changed. For this purpose, a macro, **S_CALC()**, is used:

#define S_CALC(vll,vur,wll,wur)   (((vur)-(vll)) / ((wur)-(wll)))

Although we could have used a function to calculate the scaling factors in **dowind()** and **do-**

view(), a macro can also be used to initialize **sfact[]**, as shown in the code above. **SMALL** and **BIG** are very large negative and positive (respectively) **double** constants used to initialize **wll[]** and **wur[]**.

The code for **doview()** looks the same, except that the corner points for the input window are clipped against the maximum size of the plotting surface, to ensure that the window remains inside the physical plotting surface. Corner points of the window which lie outside the physical plotting surface are set equal to the corresponding corner of the physical plotting surface. Because the code is so similar to **dowind()**, **doview()** won't be presented.

4.7.1. Exercises

1.   Implement **doview()**, using **dowind()** as a base.

4.8. doline() and doto()

doline() is responsible for reading vectors and converting them into calls on the graphics library function for drawing lines. **doline()** takes one argument: the input buffer containing the command parameters. The indices for the x co-ordinate and y co-ordinate of the plot are first converted from character to binary form, using **sscanf()**. Vectors are then read from **stdin** using **pgetv()** until a vector of length one is read, whose single element is -1. **doline()** buffers the x and y co-ordinates obtained from the input vectors, using the previous x and y co-ordinates as one endpoint for a line and the co-ordinates immediately read as the second. This allows a series of connected lines to be plotted from a series of input vectors, without requiring the vectors to be preprocessed so that each is repeated. Before the points are given to the interface routine for the graphics library, **grline()**, the line is clipped to the data window and transformed into the plotting surface co-ordinates.

doline() is complicated enough to require some pseudo-code. In particular, the buffering of co-ordinates from one iteration of the input **while** loop to the next is done with three **double** buffers, **rp[]**, for the right endpoint, **lp[]**, for the left endpoint, and **psav[]**, for saving the untransformed right endpoint from one iteration to the next. Here is the pseudo-code:

        Get the x and y indices out of the command buffer;

        Check that the indices are greater than zero;

        while(getting an input vector doesn't return **EOF**)
        {
                Check if the vector consists of -1 only and terminate loop if so;

                Check if the vector has enough elements to plot and report error if not;

                Copy the x and y co-ordinates into **rp[]**;

```
        if( this is the first iteration)
        {
             Skip plotting and copy rp[] into psav[];
        }
        else
        {
             Copy psav[] into lp[];

             Copy rp[] into psav[] for
             for the next iteration;

             Clip to data window;

             if( line is visible)
             {
                  Transform into plotting surface viewport;

                  Plot using grline();
             }
        }

    }

    return( EOF or the number of vector elements);
```

On each iteration of the loop, **rp[]** is copied into **psav[]** before clipping and transformation, so that an untransformed version is used as **lp[]** in the next iteration.

Translating the pseudo-code into C results in **doline()**:

```
/*
**********************************
doline-read vectors from stdin, clipping
    and transforming to window co-ords.
    then plotting.
**********************************
*/

doline(fin,buf)

  FILE *fin;
  char buf[];

  {
    extern double vll[],vur[];
    extern double wll[],wur[];
```

```
    extern double sfact[];
    int ix,iy,n,first_time = YES;
    int sscanf(),clip(),pgetv();
    double xbuf[MAXVEC],rp[DIM],lp[DIM],psav[DIM];

/*
  get the indices for the x and y co-ordinates
    out of the command buffer
*/

    if( sscanf(buf,"%d %d",&ix,&iy) != 2)
    {
      fprintf(stderr,
           "graphit:parameters wrong for @line:%s.\n",
           buf
           );
      return(ERR);
    }

/*
  check that the indices are legitimate
    and copy rp into psav
*/

    if( ix < 0 || iy < 0)
    {
      fprintf(stderr,
           "graphit:indices wrong for @line:%s.\n",
           buf
           );
      return(ERR);
    }

/*
  read vectors and plot
*/

    while( (n=pgetv(fin,stdout,xbuf)) != EOF)
    {

/*
  check if vector input has ended
*/
```

```
        if( n == 1 && xbuf[0] == -1)
          break;

/*
 check to be sure that the vector has enough
    elements to plot
*/

        if( n < ix || n < iy)
        {
          fprintf(stderr,
               "graphit:vector must have at least %d elements:\n",
               MAX(ix,iy)
               );
          fputv(stdout,n,xbuf,ROW);
        }
        else
        {

/*
 copy x and y co-ordinates into rp
*/

          rp[0] = xbuf[ix];
          rp[1] = xbuf[iy];

/*
 if this is the first iteration, skip plotting
*/

          if( first_time == YES)
          {
            first_time = NO;


            vcopy(DIM,rp,psav);

          }

/*
 otherwise clip, transform, and plot
*/
```

```
        else
        {

/*
copy psav into lp to get untransformed point
*/

            vcopy(DIM,psav,lp);

/*
copy rp into psav for the next time
*/

            vcopy(DIM,rp,psav);

            if( clip(lp,rp,wll,wur) == YES )
            {
              wvtrans(lp,wll,wur,vll,vur,sfact);
              wvtrans(rp,wll,wur,vll,vur,sfact);

              grline(lp,rp);
            }
          }

        }
      }

/*
issue a warning message if input ended due
  to end of file
*/

    if( n == EOF)
      fprintf(stderr,"graphit:end of file during vector input.\n");

    return(n);
}
/*
end of doline
*/
```

Error messages are generated if the command parameters are incorrect, if a vector doesn't have enough elements to satisfy the command parameters' requests for particular x and y co-ordinates, or if the end of file was reached while reading vector input. The functions **clip**() and **wvtrans**() implement the clipping and window-viewport algorithms, and are discussed later in the chapter. **MAX()** is a simple macro which takes the maximum of its two arguments, using a conditional expression:

#define MAX(a,b)   ((a) > (b) ? (a):(b))

The code for **doto()** is identical to **doline()**, except only one call need be made to **wvtrans()**, since there is only one point to transform. In addition, no point buffering is needed, since pen or cursor movement only requires one point.

4.8.1. Exercises

1.   Implement **doto()**, using **doline()** as a base.

2.   Many graphics devices have special commands which allow a series of lines to be plotted by giving only a series of endpoints, rather than repeating each endpoint twice. Modify **doline()** to accommodate this scheme.

4.9. clip()

To implement the clipping algorithm, we will need a C operator that hasn't been discussed yet: the bitwise ''and'' operator, written **&**. The bitwise ''and'' looks at its two integer operands as two patterns of bits. The result of applying the bitwise ''and'' to two integers is an integer with the same bits set to one as in both of the arguments, and all other bits set to zero.

As an example, consider taking the bitwise ''and'' of 5 and 6:

5 & 6

Translated into binary, 5 is the binary number 101 and 6 is the binary number 110. Comparing bits from left to right, the leftmost bit is 1 in both, so it is also 1 in the result. The middle and rightmost bits are not one in both numbers, so these bits are 0 in the result. Performing the bitwise logical ''and'' yields 100, or 4 in decimal. The logical ''and'' is often used to mask out bits by constructing a binary mask in which the bits to be turned off in the result are zero in the mask.

The C language also supports the bitwise ''or'' operation, |. The bitwise ''or'' works similarly, except that any bits which are one in either of the operands are ''turned on'' in the result. The bitwise ''or'' of 5 and 6:

5 | 6

would yield 111, or in 7 decimal, since, comparing bits pairwise in 101 and 110, at least one bit is ''on'' in each bit of each operand. Similar to the logical ''and,'' the logical ''or'' operation is sometimes used with a bit mask to turn particular bits on, by having those bits to be turned on in the result set to one in the mask.

The bitwise logical operators **&** and **|** should not be confused with the logical connective operators **&&** and **||**. To illustrate the difference, **4 && 2** is one, since both operands are nonzero, while **4 & 2** is zero, because the bit patterns for the two operands are 100 and 010, and neither 100 nor 010 have a one bit in common. In addition, the bitwise operators do not guarantee left to right or short circuit evaluation, while the logical connectives do.

The clipping algorithm requires that we define nine constants containing the four-bit numbers for the different regions:

```
#define M     0
#define L     1
#define R     2
#define D     4
#define U     8
#define DL    ( D | L )
#define DR    ( D | R )
#define UL    ( U | L )
#define UR    ( U | R )
```

The **M** (for "middle") code is given to points in the window, the other numbers correspond to the various quadrants around the outside of the window: **L** for "left," **R** for "right," **D** for "down," and **U** for "up." The constants for combinations are **DL** for "down left," **DR** for "down right," **UL** for "up left," and **UR** for "up right," and are formed by using the logical "or" operator to combine the codes from the two areas which form the combination. For example, "down" is 4 (100 binary) and "right" is 2 (010 binary). Taking the logical "or" for "down right" gives 110 binary, or 6 decimal, which is the code for the lower right quadrant in Fig. 4.2.

The pseudo-code also suggests a function to determine the four-bit position encoding for a point. This function, appropriately called **code()**, takes three parameters: the point to be classified, the lower left corner, and upper right corner of the window:

```
/*
*********************************
code-codes whether p is up, down, left,
    right, or in window.
*********************************
*/

int code(p,lleft,uright)

  double p[],lleft[],uright[];

{
  int c;

    c = M;

    if(p[0] < lleft[0])
      c = L;

    else if(p[0] > uright[0])
      c = R;

    if(p[1] < lleft[1])
      c = c | D;

    else if(p[1] > uright[1])
      c = c | U;

    return(c);
}
/*
end of code
*/
```

The code for **clip**() follows the pseudo-code implementation closely. The bitwise logical "and" is used to determine if the logical "and" of the four-bit codes for the two co-ordinate endpoints is zero:

```
/*
*********************************
clip-clip the line running from p1 to p2
     so it fits in the boundaries
     specified by lleft and uright.
     Return YES if any part of the
     line is visible, NO otherwise.
*********************************
*/

int clip(p1,p2,lleft,uright)

double p1[],p2[],lleft[],uright[];

{
  int c,c1,c2;
  double p[DIM];
  int code();

/*
 get four bit location codes
*/

    c1 = code(p1,lleft,uright);
    c2  = code(p2,lleft,uright);


/*
 continue until line is in window or out
*/

    while( (c1 != M) || (c2 != M) )
    {

/*
 line is outside, so reject
*/

      if( (c1 & c2) != M )
        return(NO);

/*
 set code to test
*/
```

```
            c = ( c2 == M ? c1:c2 );

/*
 point is to left
*/

        if(c == L || c == DL || c == UL)
        {
          p[1] = (p2[1]-p1[1]) *
                 ((lleft[0]-p1[0]) / (p2[0]-p1[0])) +
                 p1[1];

          p[0] = lleft[0];
        }

/*
 point is to right
*/

        else if(c == R || c == DR || c == UR)
        {
          p[1] = (p2[1]-p1[1]) *
                 ((uright[0]-p1[0]) / (p2[0]-p1[0])) +
                 p1[1];

          p[0] = uright[0];
        }

/*
 point is below
*/

        else if(c == D || c == DL || c == DR)
        {
          p[0] = (p2[0]-p1[0]) *
                 ((lleft[1]-p1[1]) / (p2[1]-p1[1])) +
                 p1[0];

          p[1] = lleft[1];
        }

/*
 point is above
*/
```

```
          else if(c == U || c == UL || c == UR)
          {
            p[0] = (p2[0]-p1[0]) *
                    ((uright[1]-p1[1]) / (p2[1]-p1[1])) +
                    p1[0];

            p[1]=uright[1];
          }

/*
 transfer into endpoint
*/

          if(c == c1)
          {
            vcopy(DIM,p,p1);
            c1 = code(p,lleft,uright);
          }

          else
          {
            vcopy(DIM,p,p2);
            c2 = code(p,lleft,uright);
          }
        }

        return(YES);
}
/*
 end of clip
*/
```

If **p1[]** and **p2[]** are the same point, **clip()** returns **NO** if the point is outside the window, and **YES** if it is inside. **clip()** can thus be used in **doto()** to determine whether or not to move the cursor, eliminating the need for a separate function.

4.10. wvtrans()

Implementation of the window-viewport algorithm is particularly straightforward, since, like the dot product, the mathematical expression need only be translated directly into code. Here is the code for **wvtrans()**, a C function which takes as parameters a point to be transformed, the corner points of the window and viewport, and the scaling factor:

```
/*
******************************
wvtrans-transform the point p in the
        window bounded by wll[]
        and wur[] into the viewport
        bounded by vll[] and vur[].
******************************
*/

wvtrans(p,wll,wur,vll,vur,sfact)

  double vll[],vur[];
  double wll[],wur[];
  double p[],sfact[];

{

    p[0] = (p[0]-wll[0]) * sfact[0] + vll[0];

    p[1] = (p[1]-wll[1]) * sfact[1] + vll[1];

    return;
}
/*
end of wvtrans
*/
```

4.10.1. Exercises

1. What happens in **clip**() and **wvtrans**() if a line runs through the corner point of a window or viewport? What happens if one of the endpoints is on a corner point?

2. Some graphics devices have the capability to do window-viewport transformation and clipping in hardware. Hardware windowing is faster, but may not be as flexible. When would software windowing be preferable to hardware windowing? If your graphics device can do hardware windowing, reimplement **doline**(), **doto**(), **do-view**(), and **dowind**() to take advantage of it.

4.11. grcolor()

grcolor() changes the color in which lines are drawn and lettering is written. We have assumed that the underlying graphics library uses small integers to designate colors on the graphics device, so **grcolor**() may need to be changed if your graphics library uses a different scheme. However, the principles of mapping the color names into a form understood by the graphics library should be portable without much difficulty.

Like **doline()**, **grcolor()** takes a character buffer with the command parameter to the **@color** command. The command parameter is a color name string, which must be matched against the color table entries to find the appropriate output code. The following implementation of **grcolor()** uses the character pointer data structure ***ctable[]** for the color table:

```
/*
*********************************************
grcolor-match input buffer color command against
        lookup table and change color.
*********************************************
*/

grcolor(buf)

  char buf[];

{

/*
 the color lookup table with codes equivalent
   to the indices of the matching string
*/

  static char *ctable[] =
  {
    "black",
    "blue",
    "green",
    "yellow",
    "orange",
    "magenta",
    "red",
    "white",
    0
  };

  int i;
  int strcmp();
  char color[MAXBUF];

/*
 get command parameter
*/
```

```
            *color = 0;
            getword(0,buf,color);

            if(*color == 0)
            {
              fprintf(stderr,
                    "graphit:no color argument to @color.\n"
                  );
              return;
            }

     /*
     look up control code in color table
     */

            for(i = 0; ctable[i] != 0; i++)
            {
              if( strcmp(ctable[i],color) == 0 )
              {
                grchclr(i);
                break;
              }
            }

            if( ctable[i] == 0 )
              fprintf(stderr,
                    "graphit:@color parameter wrong:%s.\n",
                    buf
                  );

            return;
     }
     /*
     end of grcolor
     */
```

Two checks for errors are made in **grcolor()**. If no command parameter is found in the input buffer, an error message is issued. Correspondingly, if the entire **ctable[]** array is searched and a match is not found, the user is informed that the command parameter was incorrect. The library interface routine for changing colors is assumed to have the name **grchclr()**.

Note that the pointer form of array indexing was used to obtain the value of **color[0]** in the first error check. Since **color** is a pointer to the first character in the array, ***color** dereferences the pointer, giving the character. Pointer dereferencing can be used to obtain the value at any position in an array. If **i** is the index of the desired element, then ***(x + i)** is equivalent to **x[i]**, though perhaps somewhat more clumsy to write, unless **i** is zero. **x + i** must be surrounded by parentheses; otherwise **i** is added to the value obtained from ***x**. The array form is generally preferable to the pointer form, since omitting the parentheses can easily lead to syntactically correct statements which, nevertheless, produce runtime errors.

The linear lookup scheme used in this implementation will become inefficient if the number of colors increases much. **grcolor**() must search through the entire table every time it needs to get a color near the end. The more colors, the longer the search time. As a first cut, however, the linear search algorithm is simple and easy to implement. More refined algorithms for table lookup can be implemented if the number of colors becomes so large that too much time is being spent in lookup.

4.11.1. Exercises

1.   Reimplement **grcolor**() using the **struct** array discussed in Section 4.3 for the color table.

2.   Investigate an improved algorithm for table lookup and implement it.

4.12. Some Examples

Fig. 4.3 is an example of how **graphit** can be used to compose complex pictures from simple components. The house in Fig. 4.3 was put into a separate file, and was replicated on the left and right sides of the screen by simply moving the plotting viewport. The **@file** command switches processing to the file containing the commands to plot the house. In addition, the data window for the plot on the right of Fig. 4.3 was modified to clip out the left side and top of the house, as an illustration of clipping. The arrow was also composed in a separate file and plotted by simply moving the plotting viewport, as were the boxes representing the data window and plotting viewport around the houses.

In Fig. 4.4, a sine curve was generated and the results were plotted on top of a pair of axes. The axes were composed in a separate file, and plotted first, and the sine curve was fitted on top using a separate data window to scale the curve correctly. In this case, the plotting viewports for both the sine curve and the axes were the same, since they were to appear on the same part of the screen. For plots which require more precision, the axes can be calibrated and plotted along with the data, but this method of plotting is useful when a series of quick results are needed.

graphit only supports a limited set of graphical transformations (window-viewport, clipping), but the vector-matrix tools from Chapter 2 can be used to extend this set somewhat. For example, one graphical transformation which is often useful for placing drawings without changing the data window or plotting viewport is translation of the origin. If a data point in a plot is given by (x, y), then translating the origin to the point (a, b) causes the data point to translate to $(x + a, y + b)$. The new data point will be subject to the same clipping and window-viewport transformations as the old. Translation of the origin allows you to place multiple copies of a drawing on a portion of the plotting surface without changing the plot

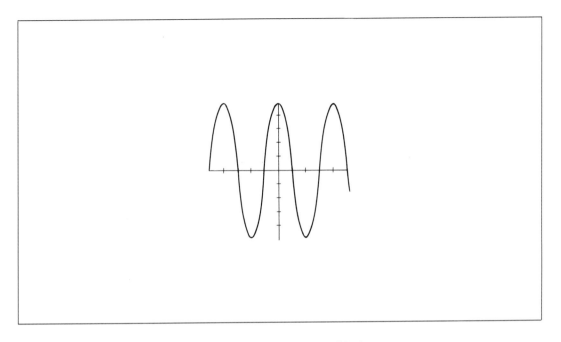

Figure 4.4. Plotting a Curve with Axes

viewport.

addm and **graphit** can be used in a pipe to achieve co-ordinate translation. The command line for the UNIX shell is:

$ **addm** < <filename> | **graphit**

where <filename> is the name of the file in which the commands for drawing the plot are contained.

An example of translation is shown in Fig. 4.5 and part of the command script to generate it is given in Fig. 4.6. The basic drawing is the arrow from Fig. 4.3, but the co-ordinates need to be duplicated instead of put in a separate file, so they can be operated upon by **addm**. The lower arrow was plotted without translation, while the upper had the origin translated to (0.1, 0.1). The plotting commands are all preceded by the pipe symbol, |. From the vector-matrix pipe protocol in Chapter 3, the pipe symbol causes **addm** to skip trying to interpret the commands, and pass them along to **graphit**. Only the co-ordinates of the new origin and of the arrow itself need be processed by **addm**.

The command entry format for translation using **addm** is not particularly convenient, since the location of the new origin must be entered before each point in the drawing. We would like to be able to enter this point only once, and have **addm** use the point for all future transformations. Modifying **addm** toward this end would be relatively simple. On the other hand, if we found that origin translation is an operation which we frequently need in plotting, a

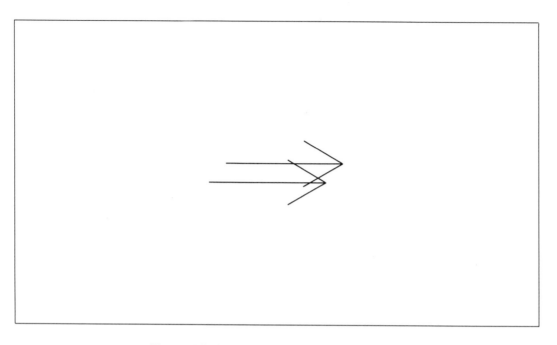

Figure 4.5. Origin Translation using addm

new command could be installed in **graphit** to perform it. If translation of the origin is a "feature" which is rarely needed, however, then adding the code to **graphit** makes little sense when a combination of **graphit** and **addm** will perform the task adequately. By using two tools together we can avoid the evils of "creeping featurism": the tendency of programs to accumulate all kinds of features as they become older. "Creeping featurism" results in gigantic programs which are difficult to maintain and use, and is the antithesis of the tools approach.

4.12.1. Exercises

1.   Modify **addm** to use the first matrix it reads as the first operand in all matrix additions. Try replotting the example. Is input any cleaner?

2.   Modify **graphit** to include an **@trans** command. The **@trans** command takes two parameters: the x and y co-ordinates of the new origin. All future data points are first translated using these points before being clipped and transformed into the plotting viewport.

3.   Another useful graphical operation is scaling. If (x, y) is a data point in a drawing, and a and b are the scaling factors in the x and y directions, respectively, then the scaled point is $(a \cdot x, b \cdot y)$. Scaling can be used to shrink or expand a drawing. How can scaling be achieved using the vector matrix tools of Chapter 2 and **graphit**?

```
|@clear
|@view 10 10 250 250
|@wind 0 0 1 1
|@line 0 1
0 0

0.2 0.5

0 0

0.8 0.5

0 0

0.6 0.65

|-1
|
|@to 0 1
0 0

0.8 0.5

|-1
|
 . . .
```

Figure 4.6. Command Script to Produce Translated Arrows

4.   A matrix is called a rotation if it has the form:

$$R = \begin{bmatrix} \cos\theta & -\sin\theta \\ \sin\theta & \cos\theta \end{bmatrix}$$

where θ is in radians. Multiplying a point $\vec{x}$ by this matrix causes the point to be rotated about the origin through an angle θ. With matrix multiplication and a rotation, a drawing can be rotated about the origin so that, for example, the arrow in Fig. 4.3 can be made to point upwards. Use the **matp** program from the exercises to Section 2.14 and **graphit** to plot rotated versions of the arrow.

4.13. Calling Foreign Language Routines from C

Many existing graphics subroutine libraries (particularly those for mainframes and minicomputers) are written in FORTRAN, Pascal, or some other language. If, however, you are working on a microcomputer, a BASIC interface is probably provided, although the actual graphics routines themselves may be written in assembly language. In any case, accessing the graphics display from C may require calling routines whose source is in another language. Foreign language calling can also be useful for incorporating existing numerical libraries into the code for the numerical tools. A library Gaussian elimination routine might be substituted for **leqsolv()** in **eqsolv**, for example.

Most modern language implementations permit foreign language calling at the source level, especially on minis and mainframes. Because microcomputer compilers for different languages are often written by different vendors, they may or may not provide compatible foreign language function-calling support, however. A general rule of thumb for foreign calling is that the more complicated the data types in the formal parameter list, the more difficult will be the interface. Foreign language global variable access is usually out of the question from the start, and must be dealt with using interface routines which pass the globals as parameters. For certain language implementations (especially on microcomputers), foreign language interfacing may require assembly language programming, or may be so difficult as to be impossible.

Some difficulties are language specific, but others are machine dependent. The way a foreign language and C are implemented on a particular machine will influence the way arguments are passed. If you are planning on trying to interface a foreign language and C, it's a good idea to get a copy of the documentation describing the implementations for both languages. A couple of quick hints, however, may reduce the amount of searching through manuals.

Most FORTRAN compilers pass all arguments by reference (there are, however, exceptions). When calling FORTRAN from C, pointers to everything must be passed. This presents no problem for arrays, which are passed by reference in C anyway, but special provisions must be made for scalars. The address operator (**&**) can be used in C to simulate passing a scalar variable by reference. For example, if **Dxlin()** is a FORTRAN subroutine which takes three arguments: the number of points in a line, a double precision array with the x co-ordinates, and a double precision array with the y co-ordinates, then **grline()** can be written:

```
/*
*****************************
grline-transfer the endpoints to x and
        y arrays and call FORTRAN
        routine Dxlin().
*****************************
*/

grline(rp,lp)

  double rp[],lp[];

{
  int npt=2;
  double x[DIM], y[DIM];

    x[0] = rp[0];
    x[1] = lp[0];
    y[0] = rp[1];
    y[1] = lp[1];

    Dxlin(&npt,x,y);     /*FORTRAN call*/

    return;
}
/*
end of grline
*/
```

The address operator is not required before the array names **x** and **y** because C passes arrays by reference anyway.

To calling a Pascal procedure it is necessary to know whether the routine is expecting a particular parameter to be VAR. If a scalar parameter is declared to be VAR in the Pascal routine, the address must be passed down using the address operator &, exactly as in FORTRAN, provided Pascal VAR parameters are implemented with call by reference. If the scalar is not a VAR parameter, then the scalar can be passed by value, as if the call were a C call.

A one-dimensional array can be passed easily if the Pascal routine is expecting it to be a VAR. No special provisions need be made, since C will pass the address of the first element, which is what the Pascal routine is expecting. However, if the array is not a VAR parameter, then there is no easy way to call the Pascal routine. The most straightforward solution is to write a Pascal interface routine that accepts the array as a VAR parameter, but then turns around and calls the library routine with the array as a nonVAR parameter. More complicated implementations, involving knowing the details of how Pascal handles arrays passed as non-VAR parameters, are highly implementation dependent.

Passing multidimensional arrays between C and Pascal is no problem beyond the VAR-nonVAR difficulty mentioned above. Both C and Pascal store arrays by row, so the Pascal routine will get the elements in exactly the order it is expecting them. With FORTRAN, however, multidimensional arrays are stored in exactly the opposite order, namely by column. Multidimensional arrays should not be passed directly between C and FORTRAN, since the FORTRAN routine would otherwise operate on a transposed version of the array.

Rather, the matrix should first be transposed before handing it to the FORTRAN routine and the result retransposed back when returning. For example, if **m[][]** is declared in C as:

double m[2][2];

then the elements are actually stored linearly in memory as:

m[0][0] m[0][1] m[1][0] m[1][1]

If **m[][]** is passed to FORTRAN, the FORTRAN routine will expect the elements to be stored like this:

m[0][0] m[1][0] m[0][1] m[1][1]

Transposing the rows and columns in the C array gives:

m[0][0] m[1][0]

m[0][1] m[1][1]

which FORTRAN will interpret correctly.

Passing any more complicated data types becomes highly dependent on the machine and implementations for the two languages. Character strings in FORTRAN are not stored as simple arrays of one-byte characters with a null byte on the end, like C character strings. FORTRAN does not support programmer-defined data types like **struct**s, and the Pascal RECORD type is often implemented completely differently from C **struct**s. Pascal pointers often have more information associated with them than the usual C pointer. The implementation manuals for both languages should be studied carefully if foreign language calls are needed with more complicated data types.

Even better, such calls can be avoided by using interface routines in C to convert the complicated data types into simpler forms (scalars or one-dimensional arrays), that can be passed easily and a foreign language interface routine can convert the data back into whatever form the library routine is expecting before calling the library routine.

Such tactics may involve lots of function calls and therefore slow down execution for a particular operation. The alternative is to study the language implementations and write code that is highly implementation specific, but which may fall apart when you try to compile it with another compiler. If execution speed is critical, then implementation-specific code for passing complicated data structures should be isolated in a routine which is labeled in big, bold letters as being a potential source of difficulties if the program is ever moved to a different compilation environment. Fortunately, most graphics and numerical libraries don't use data structures more complicated than two-dimensional arrays.

The problems associated with interfacing C to foreign language microcomputer graphics routines are similar to those discussed above for Pascal and FORTRAN on minis and main-frames. Many advanced BASICs have full subroutine capabilities, but the ground level BASIC supplied with most microcomputers has only the GOSUB statement, which makes no provision for passing parameters or local variables. Some manufacturers supplement their BASICs with keywords or pseudo-function calls that implement graphics functionality. These routines are often coded in assembly language with a BASIC language interface. If your C implementation allows it, you may be able to link to these functions. Some microcomputer C compilers even provide C-level (native language) support for graphics functionality, particularly if the graphics capabilities available are more advanced than just drawing lines and changing colors. If native language support is not available but linking with an assembly language graphics package is, C's call by value rule of parameter passing should be observed, and addresses passed when needed.

4.13.1. Exercises

1. One method of passing complicated data types between C and foreign language routines is to use a C **struct** to simulate the foreign data type. An example is the FORTRAN character string type. Some FORTRAN implementations store the number of characters in the string as a chunk of memory before the actual string. The programmer has considerable control over how memory is allocated in a **struct**, since the members are guaranteed to have addresses which increase as their declarations are read from left to right (or top to bottom). Each member of the **struct** begins on an address boundary appropriate to its type, so holes may occur if the C compiler must fill in space to accommodate a particular addressing scheme. An example would be a **char** followed by an **int**. If the **char** ends in the middle of a word, the C compiler fills in an extra byte so that the **int** begins on the next word boundary, for a machine with a two-byte word size.

    Some machines also require that identifiers be word aligned, so a **struct** with an odd number of **char** members will be rounded up to the next even number of bytes regardless. However, these machine-dependent characteristics are probably true of the foreign language data types as well. Investigate the possibility of simulating some foreign language data type using a C **struct**. Some C types not discussed in Chapter 1 which might be of use are:

    i. **long**: an integer which may be two machine words long rather than one, but may have the same size as an **int**. Check your compiler implementation for details.

ii. **short**: an integer which may have the same size as an **int**, or may be only half as long. Check your compiler implementation for details.

iii. **unsigned**: an integer which is always positive.

The members of the **struct** can be declared with these sizes so that the byte alignment of the **struct** corresponds to that of the foreign data type.

4.14. References

The field of computer graphics is much broader than the very brief and simple introduction presented in this chapter. In particular, interactive computer graphics is becoming increasingly important, beyond the filter approach presented here. Three good references on general computer graphics and algorithms are:

1. *Principles of Interactive Computer Graphics*, by William M. Newman and Robert F. Sproull, McGraw-Hill, New York, NY, 544 pp., 1979.

    A good introduction to both the hardware and software aspects of computer graphics. Contains some basic algorithms, like the clipping and window-viewport algorithms presented in this chapter.

2. *Interactive Computer Graphics*, by Wolfgang K. Giloi, Prentice-Hall, Englewood Cliffs, NJ, 352 pp., 1978.

    Contains lots of graphics algorithms, including some more advanced ones.

3. *Fundamentals of Interactive Computer Graphics*, by James D. Foley and Andries van Dam, Addison-Wesley, Reading, MA, 672 pp., 1984.

    A wide-ranging book on both hardware and software. Discusses the ACM SIG-GRAPH CORE and international GKS graphics standards.

5

FINDING OPTIMA
TO
NONLINEAR FUNCTIONS

Many phenomena of engineering and scientific interest are modeled by scalar valued functions that map one or more independent variables into a dependent variable. Often the independent variables represent quantities over which the engineer or scientist has some control, the function models some physical process, and the dependent variable is the outcome of that process. One example is the manufacture of a product. The independent variables are the amounts of raw materials necessary to produce the product and the dependent variable is the amount of product produced. The function models the manufacturing process.

Following the example further, suppose we would like to maximize production of a particular product, and we need to know what quantities of raw materials are necessary to assure maximum production. If the production process is modeled by a differentiable function, then we can use the classical techniques of differential calculus to discover the optimum. Say the function to be optimized, called the objective function, is given as:

$$f(\vec{x}) = f(x_1, x_2, \ldots, x_n)$$

Differentiating the function with respect to each independent variable and setting the resulting vector of partial derivatives (the gradient vector) to zero gives a set of n nonlinear equations, which, theoretically at least, could be solved for the set of $\vec{x}$ where the function has its maxima:

$$\nabla \vec{f}(\vec{x}) = \left(\frac{\partial f}{\partial x_1}, \frac{\partial f}{\partial x_2}, \dots, \frac{\partial f}{\partial x_n} \right) = \vec{0}$$

An important special case is when $\vec{x}$ has only one element. Since the independent variable is now a scalar, x can be written instead of $\vec{x}$ and $f(x)$ is a function of one variable. The derivative condition is:

$$\frac{df}{dx} = 0$$

In practice, if the function is nonlinear, the partial derivatives usually are also. Solving a general system of nonlinear equations analytically is not possible, unless the equations take on some special form. If the function is quadratic in all the independent variables, for example, the partial derivatives are linear and Gaussian elimination can be used to find the solution. For a general nonlinear optimization problem, however, Gaussian elimination cannot be used.

The zero condition on the partial derivatives is not enough to determine that the point is a maximum, since minima and other extrema will also satisfy the equations. Explicit second order conditions exist which can be checked to determine whether a particular point satisfying the first order conditions is a maximum. These second order conditions are complicated, however, and are rarely used in practice. The references at the end of the chapter describe them in more detail.

A simpler way is to move slightly away from the suspected maximum and evaluate the function there. If the value of the function is less for all such small perturbations, then the point must indeed be a maximum. For example, if $\vec{x}_m$ is a zero of the partial derivative equations, we can evaluate $f(\vec{x})$ at $\vec{x}_m$ and $\vec{x}_m + \vec{\varepsilon}$, where $\vec{\varepsilon}$ is a vector, the absolute value of whose elements are sufficiently small. If the following inequality holds:

$$f(\vec{x}_m) > f(\vec{x}_m + \vec{\varepsilon})$$

for all $\vec{\varepsilon}$, then $\vec{x}_m$ is, in fact, a local maximum, though it may not be the global maximum. The function may have more than one local maximum and only by examining all of them using this method can the global maximum be identified. The local maximum having the largest value of f will be the global maximum.

Of course, if the objective function is nondifferentiable then the classical methods cannot be applied. Practical problems often have objective functions which are nondifferentiable or are only piecewise differentiable. Algorithms exist for handling such cases, and we'll consider one in this chapter, called the dichotomous search algorithm, for optimizing a function of a scalar.

Many practical problems involve constraints which limit the values the independent variables may assume. An example problem might be:

    maximize: $f(\vec{x})$

    subject to: $\vec{g}(\vec{x}) \geq \vec{0}$

where $\vec{g}(\vec{x})$ is a vector of functions in $\vec{x}$ giving the constraints. This restricts the solutions geometrically to a particular region in the independent variable space. If the objective function and all the constraints are linear, then a powerful algorithm, the simplex method, exists, which will guarantee a solution, or an indication that no solution exists, for most problems. For the general case, however, there is no such algorithm. In the last section of this chapter, we'll consider a practical problem which illustrates how an unconstrained algorithm can be used to solve a problem with a constraint.

Most numerical algorithms for unconstrained and constrained optimization of differentiable functions don't use the classical calculus method directly, although they do use certain aspects of it. In addition to the dichotomous search algorithm, we'll discuss the gradient search algorithm in this chapter, which uses aspects of the classical method. The gradient search algorithm is the most fundamental method for optimizing a differentiable nonlinear objective function, and many other algorithms are modifications of it. However, when we design our software tools for solving optimization problems, we don't want to design in assumptions about particular algorithms, but rather we want to make the parts as independent as possible, so we can freely switch algorithms and objective function models without having to do massive coding. In the next section, the problem of designing problem-independent software is discussed.

5.1. Designing Problem-Independent Software

The software involved in implementing the dichotomous search and gradient search algorithms differs in a fundamental way from that presented in Chapter 2. Optimization problems invariably involve a mathematical model (the objective function) which must somehow be linked with the code implementing the algorithm and managing i/o before an executable tool can be produced. The vector-matrix tools in Chapter 2 required no special code implementing a model. Our tools for optimization in this chapter are rather more like socket wrenches than like screwdrivers, since, like a socket wrench, we first need to attach the proper piece before the tool will work. But that doesn't mean we need to recast the entire wrench every time we use it for a new job. If the supporting software is correctly written, only the model-dependent code need be rewritten, and the production of an executable image should require only that the model code be compiled and linked with the algorithm-dependent and algorithm-independent code to produce a final executable image This should also be true of code for the particular algorithm. If we want to replace the dichotomous search algorithm by another, we should not need to rewrite the entire program from top to bottom, but only the particular routines affected.

Optimization software is a good test case for designing problem-independent software, since it involves a number of different ways program parts need to interact. Practically all the vector optimization algorithms, like the gradient search algorithm, require a scalar optimization algorithm (like dichotomous search) in order to work. Using some vector operations, a function of a single variable is constructed and this function is optimized with a scalar algorithm. The result takes part in the construction of a new scalar function, if necessary, and iteration continues until some measure of the distance to the maximum is less than ε, a user-defined upper bound, or the maximum number of iterations is reached. With this strategy, a vector problem can be reduced to a scalar problem that may be easier to solve.

The constructed function of a scalar argument is usually formed by taking a ''slice'' of the objective function along an ascent direction and forming a linear combination of the current point and the ascent direction. If $\vec{x}_k$ is the point on the kth iteration through the vector algorithm and $\vec{d}$ is an ascent direction, then the scalar function is:

$$f(\ \vec{y}_k(\lambda)\ ) = f(\vec{x}_k + \lambda \vec{d})$$

where $\lambda > 0$. A vector $\vec{d}$ is an ascent direction if the following holds:

$$\frac{df(\ \vec{y}_k(\lambda)\ )}{d\lambda} \ = \ \vec{d} \cdot \nabla \vec{f}(\vec{x}_k) > 0$$

provided f is differentiable. Intuitively, an ascent direction is a direction in which the function is increasing. Optimizing the scalar function in this direction will move the focus of attention to a new point with larger value of f, which can be used as a basis for the next iteration, if necessary.

Most algorithms for optimizing scalar functions of vectors adhere to the following general pseudo-code pattern:

> Let **iter** be a user-defined limit on the maximum number of iterations to make before giving up;
>
> Let ε be a user-defined precision (an upper bound on a measure of the distance to the maximum) below which the maximum is considered found;
>
> Let **x[]** be a starting point for the optimization;
>
> **for(; iter > 0; iter--)**
> **{**
>> Calculate the measure that we are near the maximum;
>>
>> **if(** the measure is less than ε**)**
>>> **break;**
>>
>> Transform $f(\vec{x})$ into a function of a scalar argument, $f(\ \vec{y}_k(\lambda)\ )$ using **x[]** and an ascent direction, $\vec{d}$;
>>
>> Optimize this function by using a scalar method, like the dichotomous search method;
>>
>> Construct a new **x[]** based on the result of the scalar optimization;
>
> **}**

```
        if( iter <= 0)
             return(ERR);

    else
             return(OK);
```

For most scalar optimization algorithms, a similar outline is followed, except the steps involved in forming a scalar function and optimizing it are replaced by calculating a new value of **x** based on some measure of the distance to the maximum:

Let **iter** be a user-defined limit on the maximum number of iterations to make before giving up;

Let ε be a user-defined precision (an upper bound on a measure of the distance to the maximum) below which the maximum is considered found;

Let **x** be a starting point for the optimization;

```
    for( ; iter > 0; iter--)
    {
        Calculate the measure that we are near the maximum;

        if( the measure is less than ε)
             break;

        Calculate a new value for x, using the measure or by other means;
    }

        if( iter <= 0)
             return(ERR);

    else
             return(OK);
```

For some vector problems, it may be easier to use a particular scalar algorithm. Perhaps the objective function is such that a scalar method not requiring calculation of the derivative is easier to use, even though the objective function is differentiable. If acceptable accuracy can be achieved, then a combination of the gradient search method and the dichotomous search method would be a good choice. On the other hand, scalar methods using the derivative often converge to an answer faster than those which don't, so a scalar method requiring a derivative may be a better selection if model-dependent calculations are time consuming. However, if the vector optimization software is written assuming a particular scalar method, the flexibility of choosing which scalar method to use may be compromised.

It should be possible to substitute the modules for one scalar algorithm with those for another without having to touch the code for the vector algorithm at all. A single linking step should be the only additional processing required, to bind the compiled object code into a single executable image. The same property should hold for a model. If yesterday's problem required optimizing an objective function for a bridge, and today's problem is a building, only the code for the model should have to be written and compiled.

This kind of isolation from changes is achieved through careful selection of interfaces and data hiding. Any program structure that is visible through an interface is liable to perturbation should the code on either side of the interface change. If a function makes certain assumptions about the tasks it has delegated to subfunctions, then changes in those tasks will propagate back up to the calling function. Access to data is one place where functions often share access to structure. If a calling function expects data to return in a vector, and the called function is changed to return it as a matrix, the calling function must be changed as well.

For the optimization tools, one of the places where knowledge of algorithm or model dependent data might creep in and contaminate code which shouldn't know about it is with model or algorithm dependent parametric information. Algorithm dependent parameters are input data required by the algorithm to specify starting conditions, numerical accuracy, or some other property. In the above pseudo-code for a general vector optimization algorithm, ε and **iter** are such parameters. The optimization algorithms are all based on iteration and require some outside help to determine when to stop. Typically, this help is an upper limit on the number of iterations before quitting (so the program doesn't run forever) and some indication of when the current vector is sufficiently close to the maximum to quit. The latter usually takes the form of an upper bound on the derivative or vector norm of the gradient vector, below which the derivative or gradient is considered to be zero, and, therefore, the maximum is considered to have been found. For algorithms which do not require a derivative, similar stopping criteria can be formulated. Most objective function models of practical engineering or scientific interest will also have a set of parameters that need to be specified before the value of the function can be calculated.

In addition to this parametric data, the optimization routines also require a starting vector, **x[]**, like the vector-matrix tools in Chapter 2. These vectors are initial "guesses," which the algorithm then refines until sufficient accuracy has been achieved (as determined by the specified parameters) that the solution process can terminate. The optimization tools can thus also be viewed as filters, like the vector-matrix tools of Chapter 2, which read starting vectors from **stdin** and write the refined estimate to **stdout**. Within the tools, the vector currently being processed and its length can be passed between functions in the same manner as with the vector-matrix routines in Chapter 2, and the vector-matrix i/o protocol in Chapter 3 can be used to fetch starting vectors from **stdin** and write the results to **stdout**.

If **stdin** and **stdout** are being used to transmit vector information, then some other channel must be utilized to get the parametric information to the optimization tools. Since this parametric information is not likely to change during a single run of the program, and since the amount of information is typically limited to a few floating point numbers and integers, we'll have this information typed in as arguments on the command line when the program is started:

    $ prog arg1 arg2 ... argn

somewhat like the arguments on the command line for **cc**, the C compiler, and other UNIX commands. These arguments are then passed to the optimization tools when the program begins to run.

Command line arguments are, in some instances, a more effective means than prompting for communicating run time data. Typing the arguments before the program begins allows the user to carefully check them and eliminates waiting for the program to prompt, which can be frustrating if the system is slow. In addition, if the computation is really long, the program can be run batch or in the background without making special provisions for having the data entered. Finally, **stdin** and **stdout** can be used for other purposes than entering parametric data that remains constant during the program. On the other hand, the user shouldn't be required to enter long vectors on the command line, since these tend to make typing in the command slow, and can lead to confusion during command entry. Long vectors can be put into a file using a text editor and the file name can be passed to the program on the command line instead.

5.2. Overall Design for the Optimization Tools

We can divide the software for a vector optimization tool into four parts:

1. An algorithm-independent main loop, which initializes decoding of command line arguments, fetches the starting vectors from **stdin**, and writes the result to **stdout**;

2. A vector optimization module, which recognizes arguments for the particular algorithm, decodes those arguments into an appropriate data structure, and implements the vector optimization method;

3. A scalar optimization module, which recognizes arguments for the scalar algorithm, decodes the arguments, and implements the scalar method;

4. A model-dependent module, which recognizes the arguments for a particular objective function model, decodes them, and implements any model-dependent code.

The software for a scalar method follows a similar design except, of course, the vector optimization module can be omitted.

We'll treat each of these modules as a separate object, which hides its internal implementation as much as possible from the outside world, and only advertises its services through some very specific, generically written function interfaces. These interfaces should be general enough that any vector optimization module can be plugged into the main loop without having to touch a single line of source code in the main loop, and that different scalar algorithms and objective function models can be treated similarly. C's special file-scoping mechanism gives a handle on the problem by allowing us to map the module objects in the above list into files. As discussed in Chapters 2 and 4, declarations of **struct**s and **static** variables are only visible within a file, and hence they can be used to hide the implementation of the vector and scalar optimization routines. The only module structure visible outside the module is the function interface. The files corresponding to the above modules are:

1. **vmain.c** for the main loop and initialization of argument fetching (**smain.c** for the scalar tools),

2. **grad.c** for the algorithm-specific argument decoding and optimization code for the gradient search algorithm,

3. **dichot.c** for the algorithm-specific argument decoding and optimization code for the dichotomous search algorithm,

4. An appropriately named module for the model, which will, of course, differ depending on the problem.

Following our socket wrench analogy further, the function interfaces are like the square indentations on the back of the sockets, which fit neatly onto the prong of the wrench handle. If the indentation on the 8mm socket were triangular, while the one on the 10mm socket were octagonal, a different handle would be needed for using each socket. In a similar manner, by standardizing the function interfaces between module objects, we allow modules to be interchanged without having to rewrite the calling code.

After all the modules have been compiled, fitting together the pieces to obtain a working optimization tool requires a single linking step:

$ **cc -o model vmain.o grad.o dichot.o model.o **
vmio.o vector.o matrix.o c.o -lm

The linking command for a scalar tool is similar, except the name of the vector module (**grad.o** in this example) can be omitted. The UNIX shell recognizes the backslash at the end of the first line as an escape character allowing the command to continue on the next line, exactly as in the vector-matrix i/o protocol.

We'll also use a standard method of passing command line parameters to programs. Most UNIX commands follow a keyword-value system to pass parameters. The keyword (or letter) is prefixed with a minus sign (-), specifying that the following letter indicates a keyword argument. We'll stick to this convention, so that each command line argument has the following format:

-<one or more keyword letters><value>

The argument's value follows immediately behind the key letters. The key letters indicate the function of the argument. For example, the key letter **e** is used to indicate that the value is for the parameter ε in the vector and scalar optimization algorithms, and the value must be convertible to a valid C **double**. If a particular keyword letter requires more than one item, with one or more spaces between, then the argument should be enclosed in quotation marks:

-**f''param.dat log.out''**

The UNIX shell will pass the entire string between the quotation marks as part of the argument and strip off the quotation marks themselves. If the quotation marks had not been included, the shell would have passed the file name after the blank as a completely separate argument.

In the vector tools, if only one keyword letter is used, then the argument value is interpreted as being for the vector optimization module. However, since ε and **iter** are needed in both the vector module and the scalar module, the arguments for the key letters **n** (for the maximum number of iterations, **iter**) and **e** (for ε) are used to set ε and **iter** in the scalar module also, as a default. If more than one keyword letter is used, then the first letter must be either an **s** or an **m**. An **s** indicates that the rest of the argument should be decoded by the scalar algorithm module, while an **m** indicates that it should be decoded by the model module. If both an **s** and an **m** appear, then the argument's value is decoded by the model module, but is assumed to be required for the scalar part of the model. Further key letters indicate what function the argument plays. For example:

-n25 -e0.01 -se0.0001 -mp"0.5 0.5" -smp"2.0 2.0"

would set ε for the vector module to 0.01 and to 0.0001 for the scalar module, the maximum number of iterations to 25 for both modules, and would send the model module two sets of parameters (indicated by the **p**): **"0.5 0.5"** through the vector module for the vector part of the model and **"2.0 2.0"** through the scalar module. The scalar optimization tools have the same command line argument scheme, except the **s** can be omitted since all algorithmic parameters are destined for the scalar algorithm.

5.3. Localizing Algorithm-Specific Data

Although we can pass vectors and their lengths to functions in the same manner as in Chapter 2, the parameter information is more difficult, since we cannot depend on every vector optimization algorithm having the same kinds of parameters, nor can we assume all scalar algorithms have the same parameters, much less the model-dependent code. If we put specific assumptions about the parameters into the function-calling interface for the different modules, the flexibility of the code will be compromised.

We'll resolve this problem by storing the algorithm-specific parameters in **static structs**, called **vp** in the vector module and **sp**, in the scalar module. By using **static** variables, we can require the vector, scalar, and model modules to implement functions which install the algorithm- and model-specific parameters, and we will not have to build assumptions about the algorithm or model into code which doesn't need to know about it. Since C **struct** definitions are only defined within the files in which they appear (which is why the definition for the **matrix** type has to be included in every file), we can define a parameter object in **dichot.c** for holding the scalar method parameters like so:

```
struct param
{
  double eps;
  int iter;
  double d;
  double delta;
};
```

```
static struct param sp;
```

and one for holding the vector method parameters in **grad.c** like so:

```
struct param
{
  double eps;
  int iter;
};
```

```
static struct param vp;
```

and not have them conflict, as long as we *never* try to declare a **struct** object for scalar parameters in the vector module, and vice versa. A **struct** for model-specific parameters can also be defined in the model module, if necessary.

We'll describe what each of the model-specific parameters in the two **param structs** do in the next two sections. The principle should, however, be clear. Each **struct** definition and **static struct** variable is private to the module in which it is defined. Functions within the module have full access to the **struct** definition and must do all the processing associated with the parameters. Outside the module, the **struct** definition is unknown, and definitions do not conflict between modules, as long as all the data structure definitions and variables have only file scope.

5.4. The Dichotomous Search Algorithm

The dichotomous search algorithm is based on a simple "divide and conquer" strategy. It operates by assuming that a single local maximum is located somewhere within a closed interval. The objective function is evaluated on the endpoints of the interval and at two interior points. The values of the function at the two interior points are compared. If the function value at the left interior point is greater than at the right interior point, then the maximum must lie between the right interior point and the left end point. Similarly, if the function value at the right interior point is greater, then the left interior point and the right endpoint form the new interval. If the function values at the interior points are equal, then both old endpoints are discarded and the old interior points become the new endpoints. This procedure is continued until the width of the interval becomes smaller than some predefined precision, or the maximum number of iterations is exceeded. The interior points are calculated by adding and subtracting a small offset, called the distinguishability constant, to the middle point in the interval. Fig. 5.1

shows how one iteration looks.

The Achilles heal of the dichotomous search method is the requirement that a single local maximum be in the closed interval. If this is not true, the method might work or might not, but it cannot be proven that a solution will be found. In particular, if there is no local maximum within the interval, the algorithm will converge to one or the other endpoint, while, if there is more than one, the algorithm could converge to one of the maxima or to none. Nevertheless, the dichotomous search method is useful, especially for functions which have no derivative or for which the derivative is difficult to calculate analytically.

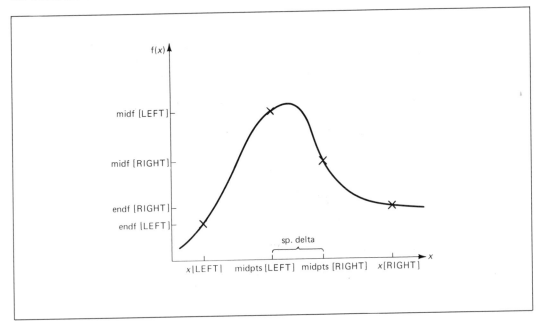

Figure 5.1. The Dichotomous Search Algorithm at Work

The pseudo-code from Section 5.1 can be specialized to the dichotomous search method as follows:

Let **sp.iter** be the maximum number of iterations;

Let **sp.eps** be the width to which the interval should be narrowed before the maximum is considered found;

Let **sp.delta** be the distinguishability constant;

Let **sp.d** be the width of the initial interval;

Let **y[0]** be the center of the initial interval;

```
x[LEFT] = y[0] - ( d / 2.0 );
x[RIGHT] = y[0] + ( d / 2.0 );
```

Get the values of the objective function at the left and right endpoints, **endf[LEFT]** and **endf[RIGHT]**;

```
for( iter = sp.iter ;
    iter > 0 && ABS(x[LEFT] - x[RIGHT]) > sp.eps;
    iter--
  )
{
    midpts[LEFT] = (x[LEFT] + x[RIGHT]) / 2.0 - sp.delta;
    midpts[RIGHT] = (x[LEFT] + x[RIGHT]) / 2.0 + sp.delta;
```

Get the function values at the midpoints, **midf[LEFT]** and **midf[RIGHT]**;

Select which endpoint to replace as follows:

```
if( midf[LEFT] > midf[RIGHT] )
{
```
    Replace **x[RIGHT]** by **midpts[RIGHT]** and **endf[RIGHT]** by **midf[RIGHT]**;
```
}
else if( midf[LEFT] < midf[RIGHT] )
{
```
    Replace **x[LEFT]** by **midpts[LEFT]** and **endf[LEFT]** by **midf[LEFT]**;
```
}
else
{
```
    Replace both endpoints and endpoint function values by the midpoints and midpoint function values;
```
}
}
```

Calculate the estimate of the maximum as the center point of the final interval;

Report an error if the precision was not achieved in the maximum number of iterations;

From the above pseudo-code, it can be seen that the value of the distinguishability constant **sp.delta** must be positive and less than $\frac{\varepsilon}{2}$, otherwise the midpoints of the final interval will be outside the interval.

In addition to the maximum number of iterations, **sp.iter**, and the precision on the final estimate of the maximum, **sp.eps**, the dichotomous search algorithm requires two other parameters: the distinguishability constant, **sp.delta**, and the width of the initial interval, **sp.d**. The distinguishability constant can be calculated from **sp.eps**, but the width of the initial interval must be passed on the command line. We'll use the flag **-d** to indicate that the argument is the initial interval width. A user can pass the width of the initial interval to a scalar tool by simply prefixing the width with **-d**, or to a vector tool with **-sd**. The **-s** in the vector tool argument indicates that it should be decoded by the scalar method module.

5.5. The Gradient Search Algorithm

For a differentiable function of several variables, the gradient vector at a particular value of the independent variable vector is perpendicular to the contours of equal value for the function, and therefore points in the direction the function is most rapidly increasing, as shown in Fig. 5.2. The gradient vector thus constitutes an ascent direction. This can be verified by checking the condition cited in Section 5.1:

$$\vec{d} \cdot \nabla \vec{f}(\vec{x}_0) = \nabla \vec{f}(\vec{x}_k) \cdot \nabla \vec{f}(\vec{x}_k) = ||\nabla \vec{f}(\vec{x}_k)||^2$$

which is always positive, so the condition holds. The gradient search algorithm uses a one-dimensional optimization method to follow the gradient until the function stops increasing, then recalculates the gradient to find a new direction. For this reason, the gradient search algorithm is sometimes called the method of steepest ascent. This procedure is iterated until either the vector norm of the gradient becomes small enough that the point qualifies as a maximum, or a limit on the number of iterations is exceeded. The procedure is illustrated graphically in Fig. 5.2.

If $\nabla \vec{f}(\vec{x}_k)$ is the gradient vector calculated at the point $\vec{x}_k$ on the kth iteration of the gradient search method, then any point along the gradient vector can be found by taking a linear combination of $\vec{x}_k$ and the gradient vector:

$$\vec{y}_k(\lambda) = \vec{x}_k + \lambda \nabla \vec{f}(\vec{x}_k)$$

where λ is a scalar. If $\lambda = 0$, $\vec{y}_k(\lambda)$ and $\vec{x}_k$ are the same point. If $\lambda > 0$, then increasing λ moves in the direction of the gradient vector, and hence in the direction of increasing f, while if $\lambda < 0$, decreasing λ moves in the direction opposite that of the gradient vector, and f is therefore decreasing.

Starting with λ set to zero, the gradient search algorithm uses a scalar optimization method to move in the direction of positive λ. With the above expression for $\vec{y}_k(\lambda)$, we can convert the objective function from being a function of a vector argument to a function of a scalar argument:

$$f(\vec{y}_k(\lambda)) = f(\vec{x}_k + \lambda \nabla \vec{f}(\vec{x}_k))$$

The maximum value of λ can now be found for this function using a scalar method, and the result used in another iteration to calculate a new value of $\vec{x}_k$, if necessary, or if the norm of the gradient is small enough, the procedure can terminate and return the calculated value of $\vec{x}_k$ as the result. If the result of the scalar optimization turns out to be zero, then $\vec{x}_k$ must already

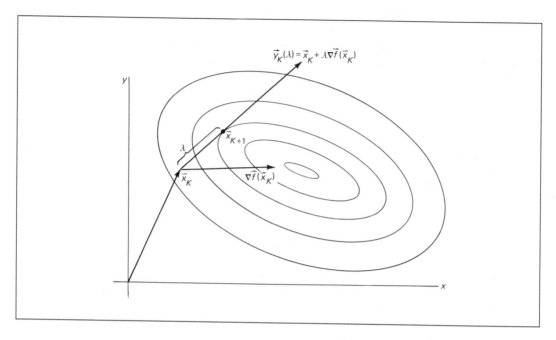

Figure 5.2. The Gradient Search Algorithm at Work

be at a local maximum, since the gradient could otherwise be followed for some distance.

For scalar algorithms which require a starting interval, like dichotomous search, starting with $\lambda = 0$ may present a problem. Since the left endpoint of the initial interval is calculated by subtracting half the initial interval width from the starting point, the left endpoint on the initial iteration of a gradient search scalar step iteration will be negative. If the scalar function of λ has only one local maximum within the interval, the second iteration of the dichotomous search method should move the left endpoint positive. But if another local maximum exists in the range $\lambda < 0$ within the interval, then the algorithm may converge to that. This problem can also occur if more than one local maximum occurs anywhere in the initial interval. As long as the scalar function is well behaved and has one local maximum within the interval, having one initial endpoint be negative should not present a problem.

The pseudo-code for the general vector optimization algorithm can be fleshed out with the details for the gradient search method as follows:

Let **vp.eps** be the upper bound on the gradient, below which the maximum should be considered found;

Let **vp.iter** be the maximum number of iterations;

Let **x[]** be the initial starting point;

for(iter = vp.iter ; iter > 0; iter--)
{

Calculate the gradient vector, **del_f[]**, for the function;

if(vector norm of **del_f[]** is less than or equal **vp.eps)**
break;

Perform the scalar optimization of $f(y_k(\lambda))$, starting it at $\lambda = 0$;

Calculate the new value of **x[]**;

}

Report an error if the precision was not achieved in the maximum number of iterations;

5.6. The main() for the Vector Tools

The main routine for the vector optimization tool has a relatively simple task. It must initialize decoding of the command line arguments, then loop, fetching starting vectors from **stdin**, calling the vector method function, and writing out the result until the end of file is encountered on **stdin**:

```
/*
*********************************************
vmain-generic main function for vector optimization.
*********************************************
*/

#include <stdio.h>
#include "c.h"

main(argc,argv)

  int argc;
  char *argv[];


{
```

```
    int len;
    double x[MAXVEC];
    int getargs(),vmethod(),pgetv();

/*
 convert command line arguments
 */

    if(getargs(argc,argv) == ERR)
    {
      vusage(argv[0]);
      exit(1);
    }

/*
 loop until no more starting points read
 */

    while( (len = pgetv(stdin,stdout,x)) != EOF)
    {

/*
 error in method occurred
 */

      if( vmethod(argv[0],len,x) == ERR)
        fprintf(stderr,
            "%s:error occurred during optimization.\n",
            argv[0]
            );

/*
 write the final result to stdout
 */

      else
        fputv(stdout,len,x,ROW);

    }
```

```
        exit(0);
}
/*
end of main
*/
```

As mentioned in Chapter 1, **main**() is a function like any other and can have arguments like any other, though the number and types of those arguments are quite restricted. In fact, the only arguments which **main**() is allowed are **argc** and **argv[]**, as shown in the code above. These arguments are used by the UNIX shell to make the command line information available to the program during execution.

A pointer to the program name is always passed down in **argv[0]**. By convention, **argc** always gives the number of command line arguments, including the program name, so **argc** is at least 1. If arguments are included on the command line after the program name, other positions in the **argv[]** data structure will contain pointers to the argument strings and the value of **argc** will equal the number of arguments plus one. After the last argument string pointer, which is in **argv[argc-1]**, **argv[argc]** is zero. As in Chapter 4, **argv[]** is declared to be an array of pointers to strings, rather than a two-dimensional array with fixed column and row sizes, because the number of argument strings can vary. In fact, in this case we have no choice but to declare **argv[]** that way, since we cannot tell how many strings the user may have typed on the command line.

If the optimization tool is started with the following command:

$ model -e0.001 -n25

then **argc** will equal 3 and **argv[]** will contain the following:

at **argv[0]** will be a pointer to the string **"model"**,

at **argv[1]** will be a pointer to the string **"-e0.001"**,

at **argv[2]** will be a pointer to the string **"-n25"**,

at **argv[3]** will be zero.

The function **getargs**() manages the parsing of the command line arguments. **getargs**() is discussed in the next section. Note that the main routine need know nothing whatsoever about the vector method's parameters, and is therefore free of any assumptions about the method used.

The main loop is pretty much the same as for the vector-matrix tools in Chapter 2. The exception is that **argv[0]** is used to print the program name in the error message, and is passed to **vusage**() and **vmethod**(), so they can report errors using the program name. Since these routines could be linked into a variety of differently named executable images, wiring the program name into their code for the error messages, as was done in Chapters 2 and 4, is not possible. **vusage**() prints out a message telling how to use the command if the user types in the arguments incorrectly. The usage message will also differ, depending on the vector and scalar optimization method, and is therefore isolated in a function which the vector optimization module has the responsibility to implement. The **vmethod()** function performs the actual optimization, and the result is written to **stdout** using **fputv()** in **main()**, if no error occurred.

The main routine for the scalar optimization tools is similar and is left as an exercise.

5.6.1. Exercises

1. Implement **main()** in **smain.c**, for the scalar optimization tools. How does it differ from the main routine for the vector tools?

5.7. getargs()

The flags before the arguments make it easy to identify them, and to type them in, since the arguments can occur in any order. **getargs()** simply needs to loop through the argument vector, checking to be sure each of the arguments begins with a -, and calling a routine to decode the argument and install it in the parameter object if so. If the argument is not marked with a dash, an error is returned:

```
/*
******************************************
getargs-decode arguments and set parameters.
******************************************
*/

int getargs(argc,argv)

  int argc;
  char *argv[];

{
  int i;
  int vpselect(),vpcheck();

/*
loop, getting arguments from argv[]
*/

    for(i = 1; i < argc; i++)
    {

/*
 check to be sure argument begins with a hyphen
*/
```

```
        if( argv[i][0] != '-' )
         return(ERR);

    /*
     now find out which argument is there and get it
    */

        if( vpselect(argv[0],argv[i][1],
                    argv[i]+2
                    ) == ERR
          )
          return(ERR);

    }

    /*
    check to be sure parameters are all OK
    */

        return(vpcheck(argv[0]));
    }
    /*
    end of getargs
    */
```

The loop starts with one as the index into the ***argv[]** array so the program name is skipped. The index form of addressing has been used rather than the pointer form, though the pointer form would have served just as well. The calculation **argv[i] + 2** in the function call to **vpselect()** skips the - and the argument character, causing only a pointer to the argument buffer to be passed to **vpselect()**. **vpselect()** does the actual decoding of the argument, installing the result in a **static** data structure in the vector module. As a result, **getargs()** is algorithm independent.

If the argument character is not a correct key letter, **vpselect()** returns **ERR**, and the usage message is printed, informing the user how to correctly use the tool. Just before **getargs()** returns, the function **vpcheck()** checks the parameters, to be sure they have all been correctly initialized, returning **ERR** if not. For example, if a user tries to initialize **eps** to a negative number, an error should be signalled. Both **vpselect()** and **vpcheck()** are passed **argv[0]**, so they can report any algorithm-specific errors using the program name.

5.8. vpselect()

The task of **vpselect()** is to decode the argument character passed as a parameter from **getargs()** and decide how to convert the argument buffer. The results of converting the argument buffer must be used to initialize the proper parameter in the vector parameter **struct vp:**

```
/*
**********************************
vpselect-decode the argument character
        and set the parameter.
**********************************
*/

int vpselect(prog,argch,argbuf)

  char *prog,argch,*argbuf;

{
  extern struct param vp;
  int mpselect(),spselect();

/*
decode the character
*/

    switch(argch)
    {

/*
precision
*/

      case 'e':
      {
        sscanf(argbuf,"%lf",&vp.eps);
        sps_eps(vp.eps);
        break;
      }

/*
number of iterations
*/

      case 'n':
      {
        sscanf(argbuf,"%d",&vp.iter);
        sps_iter(vp.iter);
        break;
      }
```

```
        /*
        argument is for scalar method
        */

            case 's':
            {
              if( spselect(prog,argbuf[0],argbuf+1) == ERR)
                return(ERR);

              break;
            }

        /*
        argument is for model
        */

            case 'm':
            {
              if( mpselect(prog,argbuf[0],argbuf+1) == ERR)
                return(ERR);

              break;
            }

        /*
        anything else is an error
        */

            default:
              return(ERR);
            }

            return(OK);
        }
        /*
        end of vpselect
        */
```

vpselect() introduces a new kind of multiway branching statement, the **switch**. A **switch** statement can be used in place of an **if-else** chain if the data upon which the decision is being made consist of an **int**. The **switch** statement is discussed further in the next section.

The **switch** statement selects what action to perform based on the value of the argument character **argch**. If the character indicates that the argument buffer, **argbuf**, is a number, the buffer is converted to binary using **sscanf()** and the result is put into the correct member of the vector parameter object, **vp**. In addition, since both the precision, **eps**, and the maximum number of iterations, **iter**, are required for the scalar algorithm, these are set by default in the scalar module as well. The scalar module has the responsibility of implementing two functions, **sps_iter()** and **sps_eps()**, which allow the vector module to do this without knowing about the definition of the scalar parameter object.

If the argument key character is **s** or **m**, then the scalar select routine (**spselect()**) or the model select routine (**mpselect()**) is called with the first character in the argument buffer as the new argument character and the rest of the buffer as the new argument buffer. These cases correspond to arguments like **-se0.0001** or **-mp"1.0 2.0"** which need to be converted by the scalar or model modules.

spselect() for the dichotomous search algorithm is essentially the same. A check must also be made for the **d** keyletter, in addition to **e** and **n**. The argument after the **d** keyletter is installed as **sp.d**, the width of the initial interval. Note that the parameters to **spselect()** are essentially the same as those to **vpselect()**, so that **spselect()** can be used alone in the scalar tools, or with a **vpselect()** for another vector tool.

5.8.1. Exercises

1.   Implement **spselect()** for the dichotomous search algorithm.

5.9. switch

The **switch** statement causes control to pass to one of several statements, depending on the value of an integer expression. The **switch** statement looks like this:

```
switch (expression)
{
  case constant1:
    statement1;

  case constant2:
    statement2;
  . . .
  case constantn:
    statementn;

  default:
    statementn+1;
}
```

After evaluation of the expression, the program's control transfers to the **case** label matching the expression, or to the **default** label if none match. The expression must evaluate to an integer and the constants must be integer constant expressions. The statements can be simple or compound; however, control flow does not automatically transfer out of the **switch** statement after the selected **case** label is finished. Unless each statement ends with a **break**, the program will continue executing into the next **case** label. The **default** label is optional and can be omitted if desired. The statements after the **default** will be executed if none of the other **case** labels apply.

The usual way to lay out a compound statement under a single **case** label is:

```
case constanti:
{
  statementi1;
  statementi2;
  statementi3;

  . . .
  statementin;
  break;
}
```

The **break** causes control to transfer out of the **switch** statement after the statements under the selected **case** label are finished executing.

In general, the **switch** statement is less useful than a multiple **if-else** statement, since the controlling conditions are quite restricted. However, it can often speed up a multiway decision, if the data upon which the decision is being made is an integer, since a **switch** statement can be compiled into code which is more efficient than a multiple **if-else** statement.

5.10. vpcheck()

After the parameters are converted by **vpselect()**, **vpcheck()** is called in **getargs()** to check that all parameters were properly initialized.

The code for **vpcheck()** first checks scalar method parameters, then the model parameters, and finally the vector method parameters:

```
/*
*************************************
vpcheck-check if all parameters correctly set.
*************************************
*/

int vpcheck(prog)

  char *prog;

{
  extern struct param vp;
  int spcheck(),mpcheck();
```

```
/*
first check the scalar parameters
*/

    if(spcheck(prog) == ERR)
      return(ERR);

/*
now check model parameters
*/

    if(mpcheck(prog) == ERR)
      return(ERR);

/*
 check vector method parameters
*/

    if( vp.eps <= 0 || vp.iter <= 0)
    {
      fprintf(stderr,
          "%s:parameters -e and -n must be positive.\n",
          prog
          );
      return(ERR);
    }

    return(OK);
}
/*
end of vpcheck
*/
```

Again, the scalar and model modules have the responsibility of implementing **spcheck**() and **mpcheck**() with the proper function-calling interface.

For the case of the gradient search algorithm, there are only two conditions to check: that the numbers given after the **-e** and **-n** arguments are positive. **vpcheck**() prints an error message if **vp.eps** or **vp.iter** were incorrectly initialized. For the dichotomous search algorithm, however, there is an additional task. The distinguishability constant, **sp.delta**, must be calculated from the value of **sp.eps**, since this algorithm-dependent parameter is not entered on the command line. **spcheck**() should also check that **sp.d** is positive and call **mpcheck**() to check the scalar model parameters.

5.10.1. Exercises

1.   Implement **spcheck()** for the dichotomous search algorithm.

5.11. vusage()

The function **vusage()** for **grad.c** and **susage()** for **dichot.c** print out a message if the user tries to start the program without giving the correct parameters. Although the exact form of the message differs depending on the tool, since the program name changes, the following message is printed by the vector optimization tool **fgrad**, presented in Section 5.15, when incorrect parameters are given:

fgrad:dichotomous search parameters:
   -e<precision>
   -d<interval size>
   -n<iterations>
   [-m<model parameters>]

fgrad:gradient search parameters:
   -e<precision>
   -n<iterations>
   [-s<scalar method parameters>]
   [-m<model parameters>]

The code for **vusage()** is simple:

```
/*
****************************************
vusage-print usage message for gradient search.
****************************************
*/

vusage(prog)

  char *prog;

{
/*
 print the usage message for the scalar method
*/

    susage(prog);
```

```
/*
print the usage message for the vector method
*/

    fprintf(stderr,
        "\n%s:gradient search parameters:\n",
        prog
        );

    fprintf(stderr,"    -e<precision>\n");
    fprintf(stderr,"    -n<iterations>\n");
    fprintf(stderr,"    [-s<scalar method parameters>]\n");
    fprintf(stderr,"    [-m<model parameters>]\n");

    return;
}
/*
end of vusage
*/
```

vusage() calls **susage()** first, so that the usage message for the scalar parameters is given first.

5.11.1. Exercises

  1.   Implement **susage()** for the dichotomous search algorithm.

5.12. vmethod() for the Gradient Search Algorithm

We now come to the routine which is the heart of the vector optimization module: **vmethod()**. By the time **vmethod()** is called, all the parameters have been installed, we have a starting vector, and we are ready to get down to the business of optimization. In basic outline, **vmethod()** follows the pseudo-code for the gradient search algorithm from Section 5.5:

```
/*
**********************************
vmethod-use the gradient search method
    to refine the maximum.
**********************************
*/

int vmethod(prog,len,x)

  int len;
  char *prog;
  double x[];

{
  extern struct param vp;
  extern double base_x[],base_d[];
  int iter, status = NO;
  int smethod(),grad();
  double l[MAXVEC],del_f[MAXVEC];
  double vnorm();

/*
 iterate, until precision achieved or maximum number of
   iterations exceeded
*/

   for( iter = vp.iter;
       iter > 0 && status == NO;
       iter--
     )
     {

/*
 get the gradient
*/

     if( grad(len,x,del_f) == ERR)
       {
        fprintf(stderr,
            "%s:error while calculating gradient.\n",
            prog
           );

       status = ERR;
       }
```

```
/*
check if the gradient is small enough to quit
*/

    else if( vnorm(len,del_f) < vp.eps)
      status = YES;
/*
use the scalar method to find maximum of f(x+lambda*del_f)
*/

    else
    {

/*
set the global values of x[] and del_f[] so the model
  module has access to them
*/

      vcopy(len,x,base_x);
      vcopy(len,del_f,base_d);

      l[0] = 0.0;

      if( smethod(prog,l) == ERR)
      {
        fprintf(stderr,
            "%s:error during scalar optimization.\n",
            prog
            );
        status = ERR;
      }

/*
calculate next x
*/

      else
        nextx(len,x,l[0]);

    }
  }
```

```
/*
 report error if ran out of iterations
 */

    if( iter <= 0)
      fprintf(stderr,
          "%s:precision %g not achieved in %d iterations.\n",
          prog,vp.eps,vp.iter
          );


    return(status == ERR ? status:OK);
}
/*
 end of vmethod
 */
```

A status variable is set to indicate why the loop terminated. The test at the end of the function determines if the iteration limit was reached, and prints an informative message if so. The value of **status** is returned as the return value of the function, in case any errors were encountered while calculating the gradient or performing the scalar optimization.

The function **grad()** returns the gradient for the objective function in the vector **del_f[]** at the point passed in the vector **x[]**. After the gradient is calculated, the vector norm of the gradient is compared to **vp.eps**, to see if it is small enough to consider the optimization finished. **vnorm()** returns the norm of the **double** vector passed as the second parameter, with the first parameter giving the vector's length. The code for **vnorm()** is a simple modification of **vdotp()** from Section 2.2 and is therefore not shown here. If the vector norm of the gradient is small enough, the status variable is set and the loop terminates.

Before the scalar algorithm is started, the gradient and current point are copied into the **static** variables **base_d[]** and **base_x[]**:

static base_x[MAXVEC],base_d[MAXVEC];

using **vcopy()**. These variables are used in **nextx()**, and are needed because the model module must be able to construct $\vec{y}_k(\lambda) = \vec{x}_k + \lambda \cdot \nabla \vec{f}(\vec{x}_k)$ on each iteration of the scalar algorithm, in order to supply the scalar algorithm with the function value. The model only has access to these variables through **nextx()**.

The actual scalar optimization is performed by **smethod()**. **smethod()** is passed (in addition to the program name, for error reporting) the starting value of λ in **l[]**. **smethod()** returns **ERR** if an error occurs during the scalar optimization calculations; otherwise, an improved estimate for **l[]** is returned in **l[0]**. This value is then used in calculating the new estimate for the maximum vector. The scalar optimization is started with the estimate **l[0] = 0**. This corresponds to saying that the present value of the independent variable vector used in the calculation $\vec{x}_k + \lambda \nabla \vec{f}(\vec{x}_k)$ is $\vec{x}_k$, i.e., the current best estimate of the maximum. An improved estimate of λ is returned from **smethod()**.

The function **nextx()** calculates the next value of **x[]**, using the **static** values **base_x[]** and **base_d[]** and the new estimate of λ returned from the scalar method in **l[0]**. The new estimate for the maximum is returned from **nextx()** in **x[]**. **nextx()** is presented in the next section. **nextx()** is used in the model module to calculate the current value of $\vec{y}_k(\lambda)$, for finding the value of the objective function.

5.12.1. Exercises

1.   The gradient search algorithm can only be used on objective functions which are differentiable. If your problem has an objective function determined by interpolation of values from a table, or some other means which precludes calculating a gradient, then the gradient search method cannot be used. One method that could be used is the cyclic co-ordinate method. The cyclic co-ordinate method is very similar to the gradient search method, except that the vectors along which line searches for a maximum are made are the co-ordinate axes rather than the gradient. As in the gradient search algorithm, the cyclic co-ordinate algorithm requires a limit on the maximum number of iterations, **iter**, and a precision against which some measure of whether the maximum was found can be compared, ε, and a starting point, $\vec{x}$. Since the objective function has no gradient, the measure of whether the maximum was found is the norm of the difference between the estimated maximum on step $k-1$ and step k:

$$||\vec{x}_k - \vec{x}_{k-1}|| < \varepsilon$$

This number is the size of the step taken by the algorithm, and termination occurs when the step size becomes so small that the algorithm is not making any significant progress. The scalar function $f(\vec{y}_k(\lambda))$ is also formed differently. Instead of using the gradient, the algorithm uses vectors parallel to the co-ordinate axes. If $\vec{d}_j$ is such a vector for the jth co-ordinate, then $\vec{d}_j$ has all zero elements, except the jth element is one. Thus $\vec{d}_1$ would have a one as the first element, $\vec{d}_2$ as the second, etc. $\vec{y}_k(\lambda)$ can therefore be defined as:

$$\vec{y}_{kj}(\lambda) = \vec{y}_{k(j-1)}(\lambda^*_{(j-1)}) + \lambda \cdot \vec{d}_j$$

The function $\vec{y}_k(\lambda)$ has been given an additional subscript j because, unlike the gradient search algorithm, it must be formed for each co-ordinate direction. If the independent variable vector has **len** elements, then the cyclic co-ordinate algorithm cycles through **len** scalar optimization steps for each vector optimization iteration, using a different $\vec{d}_j$ for each cycle, and using the maximum calculated on the previous cycle, $\vec{y}_{k(j-1)}(\lambda^*_{(j-1)})$ to form the vector argument to objective function on the current cycle. $\lambda^*_{(j-1)}$ is the result of the scalar optimization for cycle $j-1$. Here is pseudo-code for a cyclic co-ordinate **vmethod()**:

Let **vp.iter** be the maximum number of iterations;

Let **vp.eps** be the upper bound on the step size, below which the maximum is considered found;

Let **x[]** be the initial starting point, having **len** elements;

for(iter = vp.iter ; iter > 0; iter--)
{

Save the current value of **x[]** in **xold[]**, so it can be compared with the new value after the scalar optimizations;

for(j = 0; j < len; j++)
{

Let λ equal zero;

Optimize the scalar function $f(\vec{y}_{kj}(\lambda))$;

}

Construct the new point, **x[]**;

Calculate the difference between the elements of **x[]** and **xold[]**;

if(the vector norm of the difference vector <= **vp.eps**)
    break;

}

Report an error if the precision was not achieved in less than the maximum number of iterations;

Can a vector module for the cyclic co-ordinate method be written which will work with the main module, scalar module and model module function interfaces defined in the previous sections? If so, write one and test it using **dichot.c**.

2.   There are many methods for optimizing unconstrained and constrained functions with vector arguments. Investigate the references at the end of the chapter, and pick a method to implement as a numerical tool. Does it require a scalar method? Can a C module for it be written and easily linked using the interface scheme presented in this chapter? Is it possible to use the method with functions that are not differentiable? Can any of the code from **grad.c** be reused for it? Can you suggest a way of packaging the parameter handling code to reduce even further the amount of code which must be written to implement a vector algorithm?

3.   Implement a new command line argument for the vector and scalar optimization tools: **-v** for "verbose." If the **-v** flag is set, then the optimization functions print more detailed information on the optimization process to a file, whose name is

given after the **-v** flag. Such information for the gradient search method might include the independent variable vector, gradient, norm of the gradient, **l[0]**, and iteration number for each iteration. Be sure to have the information printed in neat, tabular form, identifying what each number is.

5.13. nextx()

Most of **nextx()** consists of calls to utility functions which do all the work of calculation:

```
/*
*****************************
nextx-calculate x + lambda * del_f.
*****************************
*/

nextx(len,y,lambda)

  int len;
  double y[],lambda;

{
  extern double base_x[],base_d[];

/*
 y(lambda) = lambda * del_f
*/

    vsmul(len,y,lambda,base_d);

/*
 y(lambda) = x + y(lambda)
*/

    vadd(len,y,base_x,y);

    return;
}
/*
end of nextx
*/
```

vsmul() multiplies the **double** vector in the fourth parameter by the **double** scalar in the third parameter and returns the result in the **double** vector which is the second parameter. **vadd()** adds the two **double** vectors passed as the third and fourth parameters element-wise and returns the result in the second parameter. In both cases, the length of the vector is the first parameter. These functions are relatively straightforward to implement, so the code is not presented here.

5.14. smethod() for the Dichotomous Search Algorithm

The **smethod()** function for the dichotomous search algorithm can be implemented directly from the pseudo-code given in Section 5.4:

```
/*
**************************************
smethod-use the dichotomous search method
    to refine the maximum.
**************************************
*/

int smethod(prog,x)

  char *prog;
  double x[];

{
  extern struct param sp;
  int iter;
  int model();
  double endf[2],midpts[2],midf[2];

    midf[0] = x[0];

/*
 calculate endpoints of initial interval
*/

    x[LEFT] = midf[0] - ( sp.dist / 2.0 );
    x[RIGHT] = midf[0] + ( sp.dist / 2.0 );

/*
 get endpoint function values to start
*/

    if( model(x,endf) == ERR)
      return(ERR);
```

```
/*
loop, refining maximum until iterations run out or
   interval small enough
*/

   for( iter = sp.iter;
      iter > 0 &&
        ABS(x[LEFT]- x[RIGHT]) >= sp.eps;
      iter--
    )
   {

/*
calculate two middle points
*/

      midpts[LEFT] = (x[LEFT] + x[RIGHT]) / 2.0 - sp.delta;
      midpts[RIGHT] = (x[LEFT] + x[RIGHT]) / 2.0 + sp.delta;

/*
get function values at two middle points
*/

      if( model(midpts,midf) == ERR)
        return(ERR);

/*
select which endpoint to replace
*/

      if( midf[LEFT] > midf[RIGHT] )
      {
        x[RIGHT] = midpts[RIGHT];
        endf[RIGHT] = midf[RIGHT];
      }
      else if(midf[LEFT] < midf[RIGHT] )
      {
        x[LEFT] = midpts[LEFT];
        endf[LEFT] = midf[RIGHT];
      }
      else
      {
        x[RIGHT] = midpts[RIGHT];
        endf[RIGHT] = midf[RIGHT];
        x[LEFT] = midpts[LEFT];
```

```
            endf[LEFT] = midf[RIGHT];
          }
        }

    /*
     calculate the estimate of the maximum
    */

        x[0] = (x[LEFT] + x[RIGHT]) / 2.0;

    /*
     report error if ran out of iterations
    */

      if( iter <= 0 )
        fprintf(stderr,
              "%s:precision %g not achieved in %d iterations.\n",
              prog,sp.eps,sp.iter
            );

      return(OK);
    }
    /*
    end of smethod
    */
```

The endpoints of the initial interval and values of the objective function at these points are calculated before the loop begins. The **ABS()** macro from Chapter 2 is used to calculate the distance between the endpoints. If that distance is less than the precision, **sp.eps**, the loop terminates. If the maximum number of iterations is exceeded, a warning message is printed. **smethod()** returns either **OK**, if the loop terminated, or **ERR**, if **model()** returned an error. The model module has the responsibility of implementing **model()**, which calculates the value of the model at the two points passed in the **double** vector that is the first parameter and returns the result in the **double** vector that is the second parameter.

5.14.1. Exercises

1. The dichotomous search method isn't very effective at finding maxima if the function is differentiable, since it doesn't use any information on the derivative. One method which does is the widely known Newton's method. Newton's method forms the ratio between the first and second derivatives of the function at the current value of x_k and subtracts this ratio from x_k, to form x_{k+1}. The method terminates when the absolute value of the first derivative becomes small enough to consider the maximum found. The idea behind Newton's method is, on any iteration k, to expand $\frac{df}{dx}$ in a Taylor series, set the expansion equal to zero (which is the equa-

tion for the maximum) and use the first two terms of the Taylor series approximation to derive an approximation for x_{k+1}. Proofs and more information can be found in the references at the end of the chapter. Here is the pseudo-code:

Let **sp.iter** be the maximum number of iterations;

Let **sp.eps** be an upper bound on the first derivative, below which the maximum is considered found;

Let **x[0]** be a starting point for the optimization;

for(iter = sp.iter ; iter > 0; iter--)
{
    Calculate **dfdx** and **d2fdx2**, the first and second derivatives of the objective function at **x[0]**;

    if(ABS(dfdx) < sp.eps)
        break;

    else
        x[0] = x[0] - (dfdx / d2fdx2)
}

Report an error if the precision was not achieved in less than the maximum number of iterations;

Write a scalar module for Newton's method.

2.   There are numerous algorithms for optimizing functions of one variable. Investigate some from the references at the end of the chapter and try implementing one as a scalar module. Can any of the parameter handling code from **dichot.c** be reused? Is it possible to package the parameter handling code so that the amount of code which must be rewritten to implement a scalar algorithm could be reduced even further?

5.15. fdichot.c and fgrad.c

In this section, we'll examine two model modules, one for the scalar tool using dichotomous search and one for the vector tool using gradient search. The intent here is to show how the model dependent code for two simple models looks. Most of the software for decoding parameter characters (**mpselect()**), and checking if parameter information was correctly entered (**mpcheck()**) differs little from **vpselect()** and **vpcheck()**. In addition, these models are simple enough that the analytical solution can be compared to the numerical solution, to show that the optimization tools are producing numerically correct answers. In the next section, we'll examine a more complex problem, which will require use of additional numerical tools to solve.

Consider the maximization of:

$$f(x) = 2x - x^2$$

Using the classical methods of differential calculus, we differentiate with respect to x yielding:

$$\frac{df}{dx} = 2 - 2x$$

Setting the derivative equal to zero gives the following equation for the maximum, x_m:

$$2 - 2x_m = 0$$

Solving for x_m yields:

$$x_m = 1$$

At x_m, the value of the function is:

$$f(x_m) = 1$$

We can check that this is a local maximum by perturbing x_m slightly and substituting back into the function. If the function value for the perturbed x is smaller than at x_m, for all negative and positive small perturbations, then the point $x_m = 1$ is, indeed, a local maximum.

Letting $x_p = x_m + \varepsilon = 1 + \varepsilon$:

$$f(x_p) = 1 - \varepsilon^2$$

which is always less than 1, so x_m must be a local maximum.

The model function for the dichotomous search method is relatively simple:

```
/*
******************************
model-calculate the function 2x-x**2.
******************************
*/

int model(x,f)

  double x[],f[];

  {
```

```
/*
calculate f at left and right endpoints
*/

    f[LEFT] = 2.0 * x[LEFT] - x[LEFT] * x[LEFT];
    f[RIGHT] = 2.0 * x[RIGHT] - x[RIGHT] * x[RIGHT];

    return(OK);
}
/*
end of model
*/
```

The left and right endpoints are simply calculated directly from the input vector.

To produce the final tool, we need to link together the parts using the command line:

$ cc -o fdichot smain.o dichot.o fdichot.o \
        vmio.o vector.o matrix.o c.o -lm

The tool can be run using the command:

$ fdichot -e0.0001 -n25 -d10.0

Typing in the starting point:

2.0

the result is:

1.000001185321808

which is close to the analytical result. However, if we type in the starting point:

20.0

then the result is:

15.00004407339096

which is the left endpoint of the initial interval. What happens in the second case is that the actual maximum is outside the initial interval, so the algorithm homes in on the endpoint nearest the maximum.

The model module for the gradient search tool must implement two functions in addition to **mpselect**() and **mpcheck**(): **grad**() for calculating the gradient and **model**() for calculating the objective function at the two endpoints of the interval. We'll program the function:

$$f(\vec{x}) = 2x_2 - x_2^2 - 4 + 2x_1 - x_1^2$$

The partial derivatives with respect to x_1 and x_2 form the gradient vector:

$$\frac{\partial f}{\partial x_1} = 2 - 2x_1$$

$$\frac{\partial f}{\partial x_2} = 2 - 2x_2$$

Setting both to zero and solving results in the point $\vec{x}_m = (x_{1m}, x_{2m}) = (1, 1)$. The value of f at that point is -2.

To check if $\vec{x}_m$ is indeed a local maximum, we can perturb it slightly, by adding a small constant vector, $(\varepsilon_1, \varepsilon_2)$, to $\vec{x}_m$. Evaluating the function at the perturbed value of $\vec{x}_m$ results in:

$$f(\vec{x}_p) = -2 - (\varepsilon_1^2 + \varepsilon_2^2)$$

which will always be less than -2, for any value of ε.

grad() must return the gradient vector at the point **x[]**:

```
/*
****************************
grad-check length of input vector
     and calculate gradient
****************************
*/

int grad(len,x,del_f)

 int len;
 double x[],del_f[];


{

/*
 error if input vector length incorrect
*/

    if( len != 2)
    {
```

```
        fprintf(stderr,
            "?input vector length must equal 2.\n"
            );

        return(ERR);
    }

/*
 calculate gradient
*/

    del_f[0] = 2.0 - 2.0 * x[0];

    del_f[1] = 2.0 - 2.0 * x[1];

    return(OK);
}
/*
end of fgrad
*/
```

An initial check is made to be sure the vector length is correct for the model. If it is not, an error message is printed and **ERR** is returned.

The C routine to calculate the values of the objective function at the two endpoints of the interval for the dichotomous search algorithm is slightly more complex. Since it will be called by **smethod()**, its name is **model()** and the parameters must be a **double** vector with the two interval endpoints and a **double** vector in which the two function values are returned:

```
/*
*******************************
model-calculate the function for fgrad
*******************************
*/

int model(x,f)

  double x[],f[];

{

  double vec[MAXVEC];
  double fcalc();
```

```
/*
calculate the value of f at the left endpoint
*/

    nextx(2,vec,x[LEFT]);
    f[LEFT] = fcalc(vec);

/*
 calculate the value of f at the right endpoint
*/

    nextx(2,vec,x[RIGHT]);
    f[RIGHT] = fcalc(vec);

    return(OK);
}
/*
end of model
*/
```

The first step in the calculation is to regenerate the vector $\vec{x}$ at the current iteration of the scalar method. This is done using **nextx()**, the function for calculating $\vec{y}_k(\lambda)$ which is implemented by the vector module. The value of the objective function itself is calculated by calling another function, **fcalc()**:

```
/*
************************
fcalc-calculate the function.
************************
*/

static double fcalc(x)

  double x[];

{
  double f;
  double pow();

    f = (2.0 * x[1]) - pow(x[1],2.0)  -  4.0
        + (2.0 * x[0]) - pow(x[0],2.0);

    return(f);
}
```

```
/*
end of fcalc
*/
```

Note that the function **fcalc()** was declared **static**, like the window-viewport variables in Chapter 4 and the parameter **struct**s in this chapter. In C, functions can be declared **static** too. A **static** function, like a **static** variable, is only visible within the file in which it is declared; that is, it has file scope. Outside of that file, another function with the same name could appear and there would be no conflict between the two, as long as the other function was declared **static** as well. The default scope for function names in C is **extern**. Since no function outside of **fgrad.c** will be calling **fcalc()**, there is no reason to leave it visible to the entire program.

Linking together the various parts of **fgrad**, we can make an executable tool:

**$ cc -o fgrad vmain.o grad.o dichot.o fgrad.o **
vmio.o vector.o matrix.o c.o -lm

which we can then run using the command:

$ fgrad -e0.0001 -n25 -sd10.0

Typing in the initial point:

2.0 2.0

results in the maximum being reported as:

0.9999821119308472   0.9999821119308472

which is close to the analytical result.

5.16. Estimating Materials Needs

In this section, we'll consider a real-life problem which should show how the optimization tools presented in this chapter can be used together with the vector-matrix tools from Chapter 2 to solve a problem that an engineer or scientist might encounter in daily life. Here is the problem:

*   *   *

A new company is thinking of introducing two new products, both of which require the same components to manufacture, but in different quantities. The company is able to estimate the demand for the products well, and is motivated to meet the demand closely, since overproduction causes retailers to discount and underproduction represents lost opportunities for sales.

The relationship between the number of units produced and the components from which they are assembled is linear, of the form $A\vec{x} = \vec{b}$, where $\vec{x}$ is the vector of component amounts, $\vec{b}$ is the vector of products, and A is a matrix of so-called "technological coefficients" which model the production process.

While the firm is able to estimate $\vec{b}$ fairly accurately, the major uncertainty is in the values of the technological coefficients in A. Since the process used in assembly is new, there is some room for improvement in the technological coefficients; however, improvement requires investment in the form of research and development, improvements to the existing assembly machinery, etc. From previous experience, the Manufacturing Engineering Department has provided a function of the elements of A, $f(A)$, which can be used to estimate the cost of improving the coefficients of A. Furthermore, because of technological considerations, the sum of the elements in each row of A is constrained to be one.

Summarizing the problem, the Manufacturing Engineering Department seeks the solution to the following problem:

minimize: $f(a_{11}, a_{12}, a_{21}, a_{22})$

subject to: $a_{11} + a_{12} = 1$

$a_{21} + a_{22} = 1$

and, in order to estimate the amounts of the different components:

$\vec{x} = A^{-1}\vec{b}$

* * *

Clearly, once the matrix of technological coefficients is at hand, the solution to the latter part of the problem can be found using **eqsolv** from Chapter 2. The optimization step presents several new twists, however.

The first thing to note about the optimization is that it is a minimization problem rather than a maximization problem, as we've discussed so far. But this should present no problem, since minimizing f is equivalent to maximizing $-f$ (Why?). All we need do is simply multiply the objective function by -1 and apply one of the maximization algorithms.

The constraints present a more difficult problem. In Fig. 5.3, the geometric effect of the constraint is illustrated for two of the four variables. The curves are contours of equal objective function value, and the constraint is shown as a line which cuts across the contours and limits valid solutions to the lower left corner of the first quadrant. While the true local maximum for the objective function may be outside the constrained area, as shown, the optimization algorithm must select the best choice within the constrained area. The figure illustrates that the best choice is usually a point where the function contour just touches the constraint.

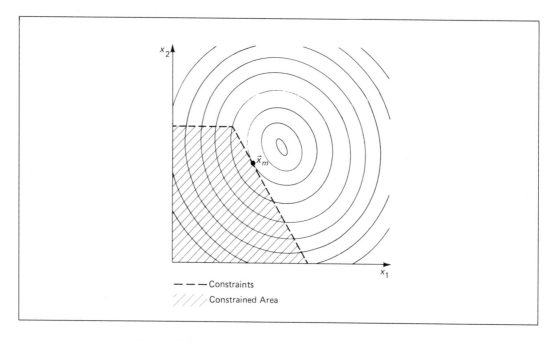

Figure 5.3. A Constrained Optimization Problem

Most practical optimization problems of any significance have constraints on the values of the variables, since the values of real world quantities rarely range over the entire real line. There are many excellent algorithms specialized for optimizing constrained problems, but, in the present case, by reformulating the objective function, we can obtain an augmented objective function which can be solved using the gradient search method. The way to do this is to make deviation from the constraints extremely costly, by incorporating the constraints into the objective function. If we multiply the square of the difference between the left- and right-hand sides of the constraints by a very large number, then add the result to the objective function:

$$\text{minimize: } f(a_{11}, a_{12}, a_{21}, a_{22}) + K(a_{11} + a_{12} - 1)^2 + K(a_{21} + a_{22} - 1)^2$$

the resulting penalty function, as it is sometimes called, can now be optimized using an unconstrained technique. If the value of K is large enough, any deviation from the constraints will cause a very large increase in the penalty function, so the problem will be forced to move very quickly to satisfy the constraints.

An additional constraint, which was unmentioned in the problem, is the sensible one that all elements of A must be greater than zero. It turns out that, if the initial value of $\mathbf{x}[]$ is near the constraint and all elements are positive, the nonnegativity constraints do not come into play, because the algorithm very quickly homes in on the constraints. However, for completeness, we could account for the nonnegativity constraints by adding a large constant to the objective function when any of the variables goes negative.

To make the problem simpler and more concrete, we'll restrict f so that there are no cross terms between the two rows. This allows us to split f into two parts, one for the first row and one for the second. We'll use the following form for f:

$$f(a_{ii}, a_{ij}) = a_{ii}^2 + a_{ij}^2 + p_1 a_{ii} + p_2 a_{ij}$$

where $i = 1$ and $j = 2$ for the first row and $i = 2$ and $j = 1$ for the second.

Summarizing the above discussion, we have now split the problem into two maximizations:

$$\text{maximize:} \quad -a_{11}^2 - a_{12}^2 - p_1 a_{11} - p_2 a_{12} - K(a_{11} + a_{12} - 1)^2$$

and:

$$\text{maximize:} \quad -a_{22}^2 - a_{21}^2 - p_2 a_{22} - p_1 a_{21} - K(a_{22} + a_{21} - 1)^2$$

The model parameters p_1, p_2, and K can be entered on the command line, since there are few of them. In solving the example, Newton's method (from the exercises to Section 5.14) was used for the scalar method instead of dichotomous search to illustrate that another scalar method could be substituted without massive reprogramming. The linking step to produce the final executable tool is:

    **$ cc -o example vmain.o grad.o newton.o example.o **
        vmio.o vector.o matrix.o c.o -lm

where **example.o** contains the compiled model module. The command lines to start the tool executing are:

    $ example -e0.001 -n50 -mK100.0 -mp"1.0 1.4"
    $ example -e0.001 -n50 -mK100.0 -mp"0.1 2.0"

with the first line being used for the first row of the matrix and the second for the second.

When the starting vector:

    0.70 0.30

was typed after the first command line, the result was

    example:precision 0.001 not achieved in 50 iterations.
    0.4818805313444843 0.5188199121521762

indicating that the final precision was not achieved for the vector method, although it was achieved on each iteration of the scalar method. This sort of behavior is typical of penalty function problems, since the portion of the penalty function contributed by the gradient may be difficult to reduce substantially if the true local maximum is located some distance outside the constraint. The penalty term then tends to dominate the optimization, quickly forcing the problem to remain within the constrained region, and the gradient of the original objective function serves only to more finely tune the location of the solution. Refinements to the penalty method which avoid this problem are discussed in the references, but if two-figure accuracy is acceptable, we can still use the result.

Using the same starting vector in the second problem results in the vector (0.04, 0.96) as the solution, so our matrix becomes:

$$A = \begin{bmatrix} 0.48 & 0.52 \\ 0.04 & 0.96 \end{bmatrix}$$

If we let:

$$\vec{b} = (323, 491)$$

then, using **eqsolv** to find $\vec{x}$, we get:

$$\vec{x} = (124, 506)$$

The point of this example is not the particular numbers involved, but to show how a problem can be solved using numerical tools instead of writing up a new program which does both the optimization and linear equation solution. In this case, the only code which really needed to be developed was the model module for the objective function f.

5.17. Call Graph for fgrad

A call graph of **fgrad** above the level of the vector-matrix i/o interface can be used to illustrate how the different modules fit together. In the following call graph, the function name has been prefixed with a capital letter indicating which module it is in (**A** for algorithmically independent, **V** for vector, **S** for scalar, and **M** for model):

```
A:main()
>>A:getargs()
>>>>V:vpselect()
>>>>S:spselect()
>>>>>>M:mpselect()
>>>>S:sps_eps()
>>>>S:sps_iter()
>>>>M:mpselect()
>>>>V:vpcheck()
>>>>>>S:spcheck()
>>>>>>>>M:mpcheck()
>>>>M:mpcheck()
>>V:vusage()
>>>>S:susage()
>>V:vmethod()
>>>>M:grad()
>>>>S:smethod()
>>>>>>M:model()
```

The emphasis in this call graph is on the intermodule calling structure, and details of the intramodule calling, as well as calls to the vector-matrix i/o module, have been suppressed.

Although there is much overhead in obtaining the parameters, once the parameters have been converted to binary form, the code for the vector and scalar methods is contained in two functions, plus whatever functions are necessary for returning information from the model. The parameter fetching routines do not slow down the inner loop of the model, and therefore any overhead is confined to the startup phase of the program. The payback in flexibility seems to be worth the cost in this case.

Many of vector and scalar algorithms take the same parameters, and so the parameter fetching and checking routines could be taken out of the algorithmic dependent files and put into a separate file, which any vector or scalar algorithm could use. Similarly, the definitions for **struct param** could easily be shared between different algorithms and models by building a standard set of **#include** files, which algorithms and models having the same parameters could simply use as is. In this manner, the amount of code necessary to bring up a new algorithm or a new model can be even further reduced.

5.18. References

There are many books on the theoretical and numerical aspects of optimization. Here are a few suggestions on where to get started, if you are interested in more information:

1. *Nonlinear Programming*, by Mokhtar S. Bazaraa and C.M. Shetty, John Wiley and Sons, New York, NY, 576 pp., 1979.

    Quite complete, covering everything from linear programming to constrained nonlinear optimization, from both the theoretical and the algorithmic side. The dichotomous search algorithm, Newton's algorithm, the cyclic co-ordinate algorithm, and the gradient search algorithm, as well as many others, are discussed.

2. *Operations Research*, by Hamdy A. Taha, Macmillan Publishing Co., New York, NY, 656 pp., 1978.

    Covers linear programming in more detail than Bazaraa and Shetty, in addition to dynamic programming and other operations research type techniques.

3. *Numerical Methods*, by Germund Dahlquist, Ake Bjoerck, and Ned Anderson, Prentice-Hall, Englewood Cliffs, NJ, 576 pp., 1974.

    Contains a chapter on optimization from the numerical standpoint, plus good material on numerical methods in general.

4. *The Engineering of Numerical Software*, by Webb Miller, Prentice-Hall, Englewood Cliffs, NJ, 176 pp., 1984.

    Has a short heuristic discussion of a minimization algorithm.

6

SOLVING
DIFFERENTIAL
EQUATIONS

Differential equations have been used to model a wide variety of real-world systems, everything from electrical circuits to alpine lake ecology. The mathematical link between these quite diverse physical systems is a postulated internal state which changes with time. The state is represented by an n variable vector of state variables, $\vec{x}(t)$. For example, the elements of the state variable vector in an ecological system model are usually the sizes of the plant and animal populations which make up the system, while the state variables in the electrical circuit are the currents and voltages in the various circuit components.

The way in which the state variables and time influence the evolution of the system is modeled as an n-element function vector, $\vec{f}(t, \vec{x}, \vec{p})$. $\vec{f}$ gives the slope of the tangent line to the solution curve, $\vec{x}(t)$, as a function of time, the state variables themselves, and sometimes a vector of parameters, $\vec{p}$, that is constant in time:

$$\frac{d\vec{x}}{dt} = \vec{f}(t, \vec{x}, \vec{p})$$

Here, $\dfrac{d\vec{x}}{dt}$ is the vector of n time derivatives for $\vec{x}$, while the independent variable t usually designates time, although t can be distance or some other quantity without changing the basic formulation. $\vec{f}(t, \vec{x}, \vec{p})$ gives the slope of the tangent line to the solution curve for fixed $\vec{x}$ and

t. The geometric relationship between $\vec{f}(t, \vec{x}, \vec{p})$ and the solution curve or trajectory is shown in Fig. 6.1 for $n = 1$.

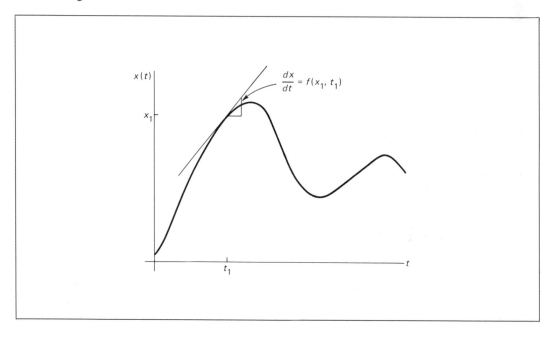

Figure 6.1. A Differential Equation as the Tangent Line to a Curve

In practice, a problem involving differential equations can take on many forms. If initial values for $\vec{x}$ and t are given, then the solution curve from the initial state to some final state may be desired. This kind of setup is often called an initial value problem. On the other hand, $\vec{x}$ may be given at two times, t_0 and t_m, and the problem may require that the evolution of the state variable vector be traced between them, if possible. Since the two values of $\vec{x}$ serve as boundaries, such a requirement is called a boundary value problem. Finally, the solution for specific values of t may not be of interest, and qualitative information on how the equilibrium state variable vector behaves as $\vec{p}$ changes when $\frac{d\vec{x}}{dt} = 0$ may be desired. Problems of this type are called bifurcation studies.

If $\vec{f}(t, \vec{x}, \vec{p})$ is a function of $\vec{x}$ only and is linear, i.e., $\vec{f}(\vec{x}, \vec{p}) = A\vec{x}$ for some $n \times n$ constant matrix A, then an exact analytical solution for an initial value problem can be found. The solution trajectory is given by a linear combination of exponentials and polynomials in t, depending on the initial conditions. Relaxing the restrictions on $\vec{f}$ somewhat, analytical solutions to the initial value problem can also be found if $\vec{f}$ is a function of t but A remains constant. In either case, the solution may not be easy to formulate if n is large; if the problem is a boundary value problem, a solution may or may not exist, depending on the boundary conditions. For a general nonlinear $\vec{f}$, however, a closed form analytical solution is impossible to generate.

In this chapter, two numerical tools for solving initial value problems are presented. As in Chapter 5, the code for the solution algorithm and problem-dependent model is isolated in separate modules, making the differential equation solvers presented here independent of any particular problem. An initial value vector, including an initial time, is input on **stdin** to start the solution process. The results of the integration are written to **stdout**, using the vector-matrix routines developed in Chapter 3, and can be postprocessed by other numerical filters if so desired. In particular, the numerical integrators can be run with **graphit**, the graphical filter discussed in Chapter 4, to convert the output into graphs. Sections 6.10 and 6.11 illustrate how **graphit** can be used with one of the numerical integration tools to plot two state variables against each other in a state plane plot, or a state variable against time, to obtain a plot of the solution trajectory.

The code modules for the two integration tools implement two different algorithms:

1.   **euler.c**-implements the Euler one-step method of integration,

2.   **rkut.c**-implements a fourth order Runge-Kutta method.

Both algorithms divide the interval between the starting independent variable value and the ending value into subintervals, and the solution on the entire interval is found by repeatedly solving the problem on one subinterval at a time. The Euler method is relatively simple, and the algorithm is discussed in more detail than the more complex Runge-Kutta algorithm. However, the accuracy of solutions produced by the Euler method is not as good as for solutions found using Runge-Kutta, for a given subinterval size. The Runge-Kutta algorithm is motivated heuristically and presented in an implementable form, similar to the discussion of the gradient search algorithm in Chapter 5. Some example model programs are given in Section 6.10, along with plots of solutions produced using **graphit**. In Section 6.11, an ecological model involving the spread of a desert into grazing land, a process called desertification, is analyzed. This problem requires the use of one of the optimization tools from Chapter 5, and again illustrates how numerical tools can be used together to solve real problems.

6.1. Data Flow Requirements of the Integration Algorithms

The data flow requirements for the integration algorithms differ very little from those for the optimization tools in Chapter 5. Both the Euler and Runge-Kutta algorithms break the independent variable interval upon which a solution is desired into **iter** subintervals of length h. The initial values for t, $\vec{x}$, the number of subintervals, and the size of the subintervals must be communicated to the program. In addition, the form of the final output must be compatible with other numerical tools, so the integration tools can be used in pipes.

The values for h, **iter**, and any model dependent parameters can be expected to remain constant throughout a calculation and are therefore candidates for command line arguments. We'll prefix the parameter h with the flag **-h** and **iter** with the flag **-n**. As in Chapter 5, parameter-like information required by the model dependent code in the model module is prefixed with the flag **-m**.

In addition to these command line arguments, however, we'll also include one for selecting which algorithm to use. Unlike the optimization algorithms of Chapter 5, the numerical integration algorithms all use the same parameters. This means that we can link the code for both integration algorithms into the same program, and let the user select which algorithm to use at run time rather than at link time. The additional flexibility of this approach allows the use of a simple algorithm which executes more quickly, like the Euler algorithm, to get a prototype solution, then a switch to a more complicated but slower executing algorithm, like the Runge-Kutta algorithm, for a more accurate final answer, all without having to relink the program. To differentiate the two, the user must supply either an **-e**, for the Euler algorithm, or an **-r**, for the Runge-Kutta algorithm, on the command line.

Input to the integration tools consists of vectors, the first elements of which are the initial values for t with the remaining elements being the initial values for the state variable vector $\vec{x}$. After the integration, the resulting vector is written to **stdout** in the same format. Arranging the data flow in this manner allows the integration tools to be interfaced with other numerical tools and **graphit**. For example, say we've built an integration tool called **model1** with four state variables, and we'd like to plot the second state variable against the fourth, in a state plane plot. The UNIX shell command:

$$\text{\$ model1 -n20 -h0.01 -e | graphit}$$

starts the integration using the Euler method.

After the command line has been typed, the following input sequence produces the desired output graph:

```
| @line 2 4
<initial values of t and x>

| -1
|
```

The first line of input is passed through **model1** and on out to **graphit**, where it is interpreted as a request to draw a line using the vectors which follow. Note that the indices of the vector elements to be used in the graph are exactly the same as the numbers of the desired state variable vector elements. The first element in the output vectors from **model1** will be the time (at input vector position zero) and not the first element of the state variable vector, so we don't have to subtract one to find the state variable when specifying which indices to use for the **@line** command.

The following vector is used as the initial value for an integration, and will cause a sequence of time and state variable vectors to be written to **graphit**, where they will be transformed into a graph. Finally, the last two lines cause the vector with -1 as its single element to be passed through **model1** and the model to **graphit**, terminating the **@line** command.

6.2. Parameter Handling and Function Pointers

As mentioned in the previous section, the two numerical integration algorithms both have the same parameters: h, the size of the subintervals in the independent variable, and **iter**, the number of subintervals constituting the interval over which the integration is performed. We can therefore use the same **struct** to hold the parameters for both, and we'll put the **struct** definition in the file **paramdef.h**, which can be included in **main.c**, **euler.c**, **rkut.c**, and any other modules which need to know about it.

An additional parameter is the algorithm itself, which is selected by the user at run time. If we stick to the language tools we already have, there are two ways we could implement this selection. An **if-else** chain (as in **graphit**) could be used to pick the appropriate function or a **switch** statement could select the function implementing the particular algorithm by testing the value of the command line argument. For both of these, the argument character could be stored in the parameter **struct** at the time the command line arguments are parsed and later used to make the choice.

Rather than going with one of these solutions, we'll try something different. In C, a function call itself can be parameterized, so that the function which is actually called depends on the value of a variable, rather than the name of a function hard wired into the calling statement at the time the program was written. This is similar to the case in FORTRAN and Pascal, where a function can be passed as a parameter to another function, subroutine, or procedure. Within the body of a called function having a function valued formal parameter, the formal parameter name can be used in a function call statement, so that, when the actual parameter is substituted, the function called depends on the value of the actual parameter.

Something similar can be done in C. For example, in the following code, **doop()** is a function with a formal parameter that is a function:

```
int doop(len,x,op)

  int len;
  double x[];
  double (*op)();

{
  double r;

  r = (*op)(len,x);

  . . .

}
```

The calling code for **doop()** simply requires that the function name be used as the last actual parameter:

```
double norm();

    . . .

if( doop(len,x,norm) == ERR)
    return(ERR);

    . . .
```

The declaration for the **op** formal parameter in **doop()** requires some explanation. C implements such variable functions as function pointers. The declaration says that **op** is a pointer (***op**) to a function (**()**). When the function is called, the variable containing the function pointer is dereferenced using the * operator, just as if the variable had contained a data value. The function passed as the actual parameter will be called when the call statement is executed. Note that this declaration is different from:

```
double *op();
```

which declares **op** to be a function returning a pointer to a **double**. Furthermore, if a function name is used as the actual parameter in a function call list, it is automatically converted into a function pointer, so there is no need to use the address operator.

C restricts the operations which can be applied to function pointers and function names; nevertheless, function pointers are considerably more flexible than function parameter arguments in FORTRAN or Pascal. In particular, a function name may also be assigned to an appropriately declared variable or array element (in which case the address of the function is deposited in the variable), in addition to being passed as a parameter to another function, as illustrated above. No other operations are permitted, however.

Using a function pointer called **integrate**, we can now include a member for the parameter **struct** which holds a pointer to the actual function to call:

```
struct param
{
  int iter;
  double h;
  int (*integrate)();
};
```

During command line argument parsing, we can assign either the function **euler()** or **rkut()** to **integrate**, depending on which algorithm the user has selected. This eliminates the need for an **if-else** chain or a **switch** statement when the function is actually called.

6.3. The main()

The **main**() for the integration tools is very similar to the main function for the optimization tools from Chapter 5, although it is structurally different because the main module, **main.c**, handles all the argument parsing and error reporting, rather than delegating these tasks to the particular algorithm module, as was the case for the optimization tools. Here is the code for **main**():

```
/*
the minimum number of elements in a
  time + state variable vector
*/

#define MIN_SVEC        2

/*
the parameter struct definition
*/

#include "paramdef.h"

/*
********************************
main-get the command line arguments
      and a starting vector for the
      differential equation solver,
      then call the integration
      function.
********************************
*/

main(argc,argv)

  int argc;
  char *argv[];

{
  struct param p;
  int len = 0;
  int getargs(), pgetv(), euler();
  double x[MAXVEC];
```

```
/*
 initialize p
*/

    p.iter = 0;
    p.h = 0.0;
    p.integrate = euler;

/*
get the command line arguments
*/

    if( getargs(argc,argv,&p) == ERR)
    {
      fprintf(stderr,"%s:parameters:\n",argv[0]);
      fprintf(stderr," -n<iterations>\n");
      fprintf(stderr," -h<step size>\n");
      fprintf(stderr," <-e | -r>\n");
      fprintf(stderr," {-m<model parameters>}\n");
      exit(1);
    }

/*
get the starting independent variable and
  the starting state variable vector
  from stdin
*/

  while( (len = pgetv(stdin,stdout,x)) != EOF)
  {

/*
check to be sure that there are at least
  MIN_SVEC elements in the vector, the
  starting time at x[0] and the starting
  state variables at x[1] through
  x[nvec-1]
*/

    if( len < MIN_SVEC)
      fprintf(stderr,
          "%s:need time and one independent variable.\n",
          argv[0]
          );
```

```
        /*
        call the integration function
        */

            else if( (*p.integrate)(argv[0],len,x,&p) == ERR)
              fprintf(stderr,
                  "%s:error while integrating.\n",
                  argv[0]
                );

        }

        exit(0);
      }
      /*
      end of main
      */
```

The integration needs at least two elements, a starting value for t in $x[0]$ and one for a scalar x in $x[1]$, which is why a check is made after the starting vector is input. The integration cannot proceed without at least two elements. The minimum number of elements allowed in a time + state variable vector is defined by the constant **MIN_SVEC**. We assume that the model function does any further checking for the proper number of state variables, and returns **ERR** through the integration function if too few are entered.

Command line parsing and error detection is done in **getargs()**, and the usual parameter message is written to **stderr** if any errors are detected. If no errors occurred in the argument parsing, **getargs()** installs the proper function in **p.integrate** and the indirect function call is done using pointer dereferencing in **main()**, as shown.

In the initialization section, notice that a pointer to **euler()** is assigned to the variable **p.integrate** rather than assigning the result of calling **euler()** because the function name is not followed by parentheses. Initializing **p.integrate** to **euler** makes the Euler algorithm the default algorithm for the integration tools. If the user makes no choice, the Euler algorithm is used, sparing the need to specify the algorithm with a command line argument.

6.4. getargs()

The **getargs()** routine differs little in structure from **vpselect()** in Chapter 5. Since no algorithm-specific parameters need be accommodated, iteration through the command line arguments and conversion of the parameter values can be isolated in one routine:

```
/*
***************************
getargs-decode argument flags.
***************************
*/

int getargs(argc,argv,p)

  struct param *p;
  int argc;
  char *argv[];

{
  extern int euler(),rkut();
  int sscanf(), mpgetargs(), pcheck();
  char *prog;

/*
 save program name
*/

   prog = argv[0];

/*
 decode argument vector
*/

   for( --argc, ++argv;
      argc > 0;
      --argc, ++argv
    )
    {

/*
 check to be sure argument begins with a hyphen
*/

    if( *argv[0]++ != '-' )
     return(ERR);

/*
 now find out which argument is there and get it
*/
```

```
      switch(*argv[0]++)
      {

/*
 maximum number of iterations
*/

          case 'n':
          {
            if( sscanf(argv[0],"%d",&(p->iter)) != 1)
              return(ERR);
            break;
          }

/*
 subinterval size
*/

          case 'h':
          {
            if( sscanf(argv[0],"%lf",&(p->h)) != 1)
              return(ERR);
            break;
          }

/*
 use the euler method
*/

          case 'e':
          {
            p->integrate = euler;
            break;
          }

/*
 use the runge-kutta method
*/

          case 'r':
          {
            p->integrate = rkut;
            break;
          }
```

```
/*
argument is for model
*/

        case 'm':
        {
          if( mpgetargs(prog,*argv[0],argv[0]+1) == ERR)
            return(ERR);
          break;
        }

/*
anything else is an error
*/

        default:
          return(ERR);
        }
      }

/*
check if all the parameter values are correct
*/

    if( pcheck(prog,p) == ERR)
      return(ERR);

    return(OK);
}
/*
end of getargs
*/
```

Note that **argv** is being used as both a pointer to an array of character pointers and as an array name. In the **for** statement, the pointer **argv** is autoincremented, so that it points to the next character pointer in the array. On the first iteration, this skips the program name (which is why the program name is first saved, for error messages), and on subsequent iterations, it moves **argv** to a new string requiring processing. In the **if** and **switch** statements, **argv** is used as an array name, and the character pointer within it is dereferenced with the * operator to test its value. After dereferencing, the character pointer is autoincremented by the postfix ++, so that it points to the next character in the string, which then becomes the current character under examination.

Model-dependent argument parsing is handled by **mpgetargs()**, which performs the same function as **mpselect()** in Chapter 5. The proper integration function is also assigned to **p.integrate** by branches of the **switch** statement. Finally, the function **pcheck()** checks whether the parameters have all correctly been initialized. **pcheck()** has the same structure as **vpcheck()** from Chapter 5.

6.5. The Euler Algorithm

The Euler algorithm was proposed by the Swiss mathematician Leonhard Euler in the 18th century as a solution for otherwise intractable initial value problems. The basic idea is to divide the interval between the initial value and final value of the independent variable into **iter** subintervals of length h. If the subintervals are small enough, the derivative, $\frac{d\vec{x}}{dt}$ in the differential equation at the beginning of the chapter can be replaced by a ratio of finite differences:

$$\frac{\vec{x}_{k+1} - \vec{x}_k}{h} = \vec{f}(t_0 + kh, \vec{x}_k, \vec{p})$$

for $k = 1, 2, ..., $**iter**. Multiplying both sides of the equation by h and adding $\vec{x}_k$ to both sides produces the recurrence relation:

$$\vec{x}_{k+1} = \vec{x}_k + h\vec{f}(t_0 + kh, \vec{x}_k, \vec{p})$$

Given an initial value for $\vec{x}$ and t, and a subdivision of the time interval, this relation can be iteratively solved for an approximation to $\vec{x}(t)$ on the interval by plugging values for $\vec{x}_k$ and h into the right-hand side.

The problem with the Euler method is that the accuracy of the solution depends upon the size of the subintervals. This dependency is a property shared with most numerical integration

t	$h = 0.1$		$h = 0.01$	
	x	error	x	error
0.0	1.000	0.00	1.000	0.00
0.1	1.100	5.17×10^{-3}	1.104	1.17×10^{-3}
0.2	1.210	1.14×10^{-2}	1.220	1.40×10^{-3}
0.3	1.331	1.88×10^{-2}	1.347	2.85×10^{-3}
0.4	1.464	2.78×10^{-2}	1.488	3.82×10^{-3}
0.5	1.610	3.87×10^{-2}	1.644	4.72×10^{-3}
0.6	1.771	5.11×10^{-2}	1.816	6.11×10^{-3}
0.7	1.948	6.57×10^{-2}	2.006	7.75×10^{-3}
0.8	2.143	8.25×10^{-2}	2.216	9.54×10^{-3}
0.9	2.357	1.03×10^{-1}	2.448	1.16×10^{-2}
1.0	2.593	1.25×10^{-1}	2.704	1.42×10^{-2}

Table 6.1. The Accuracy of the Euler Method for Two Values of h.

methods that do not adjust the size of their subintervals based upon the estimated numerical error, but the improvement obtained by using a smaller step size increases rather slowly for the Euler method. In particular, it can be shown that the error obtained using Euler's method is proportional to h, so the accuracy of an estimate increases only linearly as the step size is decreased. As a result, the step size might need to be decreased to the point where the computational cost is too high to get acceptable accuracy.

Table 6.1 compares the estimated solution of the scalar differential equation:

$$\frac{dx}{dt} = x$$

using the Euler algorithm for $h = 0.1$ and $h = 0.01$, to the analytical solution. The initial conditions for the solution were $x(0) = 1$ and $t_0 = 0$. The analytical solution is given by:

$$x(t) = e^t$$

Although the error does decrease somewhat as the step size decreases, the per step error in the numerically estimated solution for $h = 0.01$ accumulates until, by the end of the interval, the solution is not even accurate to two decimal places. Later, in Section 6.7, the same comparison is made for a solution obtained with the Runge-Kutta method.

In pseudo-code, the Euler algorithm can be outlined as:

Let **p->iter** be the maximum number of iterations;

Let **p->h** be the size of the independent variable subintervals, into which the independent variable interval is divided;

Let **x[]** be the initial time + state variable vector;

Let **x0** be the initial value of time;

Write out the initial time + state variable vector;

for(iter = 1; iter <= p->iter; iter++)
{
    Find the value of the model function $\vec{f}(t, \vec{x}, \vec{p})$ and put it in **f[]**;

    Calculate the new value of **x[]**, using the following recurrence relation for each element, **i > 0**, of **x[]**:

        x[i] = x[i] + p->h * f[i];

    x[0] = x0 + iter * p->h;

    Write out the new value of **x[]**;

}

```
        return(OK);
```

6.6. euler()

The integration routine for the Euler algorithm is a straightforward implementation of the algorithm in Section 6.5:

```
/*
*********************************
euler-use the Euler method to integrate
    the model.
*********************************
*/

int euler(prog,len,x,p)

  struct param *p;
  int len;
  double x[];
  char *prog;

{
  int nstate,iter,status=OK;
  int mdiffeq();
  double f[MAXVEC],x0,*x1;

/*
save the initial time, calculate the address
  of the second element in the state vector
  and the number of elements
*/

    x0 = x[0];
    x1 = x + 1;
    nstate = len - 1;

/*
write out the initial vector
*/

    fputv(stdout,len,x,ROW);
```

```
/*
loop, until the full number of iterations has occurred
*/

    for( iter = 1; iter <= p->iter; iter++)
    {

/*
get the equation system value
*/

      if( (status=mdiffeq(prog,len,x,f)) != OK)
        return(status);

/*
do the integration
*/

      vsmul(nstate,f,p->h,f);
      vadd(nstate,x1,x1,f);

/*
calculate a new value for time
*/

      x[0] = x0 + iter * p->h;

/*
write out the result of the integration
*/

      fputv(stdout,len,x,ROW);

    }

    return(OK);
}
/*
end of euler
*/
```

The C function implementing the differential equation system's right-hand side is called
mdiffeq(). If some error occurs in the calculation of this function, **euler()** returns immediately,
passing back the return status.

The number of state variables and the address of the second element in the array are calculated before the integration loop, rather than doing the operations every time they are needed. The two utility routines **vsmul**() and **vadd**() handily do the scalar multiplication and addition needed to complete the integration. The pointer form of array addressing is used to pass a pointer to the second element in **x[]** as the beginning element for the state variable vector in the calls to **vadd**(). This allows **vadd**() to operate only on the state variable information and avoid modifying the value of time in the first element of **x[]**, but avoids having to copy the vector.

Afterwards, the new value for time is calculated and put into **x[0]**. Notice that the new value for time is not simply calculated by adding **p->h** to the old value, but rather by multiplying **p->h** by the number of iterations, and adding that to the initial value of time. This avoids accumulating errors resulting from the truncation which inevitably accompanies floating point arithmetic, a topic which was briefly discussed in Chapter 1.

6.6.1. Exercises

1. One way of increasing the accuracy of the estimates from the Euler method while only increasing the number of function evaluations on the interval by two is to have the method adjust the step size based on an estimate of the local error. The local error is the difference between the actual value of state variable vector $\vec{x}$ and the numerically calculated value. To estimate the local error, the following procedure is used. The equations are first integrated as usual from t_k to t_k+h_k, then integrated from t_k to an intermediate point at $t_k+\dfrac{h_k}{2}$, and from $t_k+\dfrac{h_k}{2}$ to t_k+h_k. Calling the first estimate $\vec{x}_{k+1}$ and the second $\vec{x}^*_{k+1}$, the local error is approximately:

$$e_k = \frac{||\ \vec{x}_{k+1} - \vec{x}^*_{k+1}\ ||}{2^N - 1}$$

where N is the order of the method. In the case of the Euler method, $N = 1$. There are a number of possible approaches to calculating the size of the next step. One which has some theoretical basis and seems to work well is to let h_{k+1} be some percentage of h_k, weighed by the error. In particular:

$$h_{k+1} = \left[\frac{\varepsilon h_k^{\,p+1}}{e_k}\right]^{\frac{1}{p}}$$

with $\varepsilon < 1$ and p selected on the basis of experience. Incorporate a variable step size scheme into **euler**() and experiment with it on some problems. Are the results any more accurate? Compare the number of function evaluations with the Runge-Kutta method.

6.7. The Runge-Kutta Algorithm

Runge-Kutta techniques are some of the most popular methods for integrating reasonably well-behaved systems of differential equations. Partly because no information, other than $\vec{x}_k$ and t_k, need be saved from one iteration to the next, and partly because the accuracy increases rapidly as the step size decreases, the Runge-Kutta techniques are computationally efficient in comparison with more complicated algorithms and give more accurate estimates in comparison with simpler methods, like the Euler method. The one drawback is that multiple evaluations of $\vec{f}$ are required on each iteration, rather than just one evaluation, as in the Euler method, but the increase in accuracy is often worth the extra computational cost.

There is really not a single Runge-Kutta algorithm, but rather a family of techniques which differ in the number of functional evaluations required per iteration and the accuracy of the estimates for $\vec{x}_k$ with a given step size. Higher order techniques require more functional evaluations but also give better accuracy. We'll use a fourth order technique, requiring four functional evaluations per iteration, and giving an accuracy proportional to h^5. In general, the accuracy of an Nth order Runge-Kutta technique is proportional to h^{N+1}, so the error decreases much faster as the step size drops than in the Euler method.

The recurrence relation for the fourth order Runge-Kutta algorithm is given by the formulas:

$$\vec{a}_1 = h\vec{f}(t_n, \vec{x}_n)$$

$$\vec{a}_2 = h\vec{f}(t_n + \frac{h}{2}, \vec{x}_n + \frac{\vec{a}_1}{2})$$

$$\vec{a}_3 = h\vec{f}(t_n + \frac{h}{2}, \vec{x}_n + \frac{\vec{a}_2}{2})$$

$$\vec{a}_4 = h\vec{f}(t_n + h, \vec{x}_n + \vec{a}_3)$$

$$\vec{x}_{n+1} = \vec{x}_n + \frac{1}{6}(\vec{a}_1 + 2\vec{a}_2 + 2\vec{a}_3 + \vec{a}_4)$$

The constant vectors $\vec{a}_i$ in the Runge-Kutta formulas are derived by evaluating $\vec{f}$ at selected points within the interval $(t_k, t_k + h)$. These points are chosen from a consideration of a Taylor series expansion for $\vec{f}$. In fact, the recursion relation for the Euler algorithm can be considered a truncation of the Taylor series expansion after the first term. The Runge-Kutta formulas extend the Taylor expansion further, and the constant vectors $\vec{a}_i$ are estimates for the higher terms in the series. As more terms are added to the series, the error in the series estimate decreases. Consequently, the accuracy of the resulting numerical estimate increases, which explains why the Runge-Kutta algorithm gives more accurate estimates than the Euler algorithm. References at the end of the chapter discuss how to derive these formulas, for those who are interested in the theoretical aspects of numerical integration. The derivation is straightforward but requires some understanding of how to manipulate matching Taylor series.

In Table 6.2, the results of running the Runge-Kutta algorithm on the same problem as in Table 6.1 are shown. Table 6.2 also shows the calculated error between the numerically estimated solution and the analytical solution. Comparing the calculated error in Tables 6.1

and 6.2, we can see that the Runge-Kutta algorithm gives much better accuracy than the Euler method for a given value of h. In Table 6.2, the estimates for x in the first four entries are the same up to five figures, showing that excellent accuracy is possible even with $h = 0.1$. For both the 0.1 and 0.01 step sizes, the error at the beginning of interval is about h^5, but towards the end accumulated error decreases the accuracy below the theoretically attainable local error.

t	$h = 0.1$		$h = 0.01$	
	x	error	x	error
0.0	1.00000000	0.00	1.00000000	0.00
0.1	1.10517083	8.47×10^{-8}	1.10517091	9.13×10^{-12}
0.2	1.22140257	1.87×10^{-7}	1.22140275	2.01×10^{-11}
0.3	1.34985849	3.10×10^{-7}	1.34985880	3.34×10^{-11}
0.4	1.49182424	4.57×10^{-7}	1.49182469	4.93×10^{-11}
0.5	1.64872063	6.32×10^{-7}	1.64872127	6.81×10^{-11}
0.6	1.82211796	8.38×10^{-7}	1.82211880	9.03×10^{-11}
0.7	2.01375162	1.08×10^{-6}	2.01375270	1.16×10^{-10}
0.8	2.22553956	1.36×10^{-6}	2.22554092	1.47×10^{-10}
0.9	2.45960141	1.69×10^{-6}	2.45960311	1.82×10^{-10}
1.0	2.71827974	2.08×10^{-6}	2.71828182	2.24×10^{-10}

Table 6.2. The Accuracy of the Runge-Kutta Method for Two Values of h.

The pseudo-code for the Runge-Kutta algorithm differs from that for the Euler algorithm in that the recurrence relation is more complex:

Let **p->iter** be the maximum number of iterations;

Let **p->h** be the size of the independent variable subintervals, into which the independent variable interval is divided;

Let **a[4][MAXVEC]** be the matrix of constants in the recurrence relation;

Let **x[]** be the initial time + state variable vector;

Let **x0** be the initial value of time;

Write out the initial time + state variable vector;

for(iter = 1; iter <= p->iter; iter++)
{
    Calculate the value of the constant matrix **a[][]**;

Calculate the new value of **x[]**, using the following recurrence relation for each element, **i > 0**, of **x[]**:

x[i] = x[i] +
       (1.0 / 6.0) * (a[0][i] + 2.0 * a[1][i] + 2.0 * a[2][i] + a[3][i]);

x[0] = x0 + iter * p->h;

Write out the new value of **x[]**;
  }

    return(OK);

6.8. rkut()

The structure of **rkut()** is very similar to **euler()**, except that an inner loop calculates the constants $\vec{a}_i$:

```
/*
define multiplication constants for the 4th order
  Runge-Kutta formula
*/


#define NUMERATOR        1.0
#define DENOMINATOR      6.0
#define FACTOR           2.0
#define A_FACTOR         0.5


/*
**************************************
rkut-integrate the model function using a 4th
    order Runge-Kutta formula.
**************************************
*/

int rkut(prog,len,x,p)

  struct param *p;
  int len;
  double x[];
  char *prog;
```

```
{
  int nstate,iter,i,status=OK;
  int mdiffeq();
  double f[MAXVEC],a[4][MAXVEC];
  double x0,ratio,*x1;

/*
save the initial time, calculate the address
  of the second element in the state vector
  and the number of elements
  and the address of the second element
*/

  x0 = x[0];
  x1 = x + 1;
  nstate = len - 1;

/*
calculate the ratio for the recursion relation
*/

  ratio = NUMERATOR / DENOMINATOR;


/*
write out the initial vector
*/

  fputv(stdout,len,x,ROW);

/*
loop, until the full number of iterations has occurred
*/

  for( iter = 1; iter <= p->iter; iter++)
  {

/*
get the values of a[][]
*/
```

```
        for( i = 0; i < 4; i++)
        {

/*
a1 requires no preparation, just use x in the function call
*/


/*
a2 and a3 require multiplying by a scalar and adding
*/

          if( i >= 1 && i <= 2)
          {
            vsmul(nstate,a[i],A_FACTOR,a[i-1]);
            vadd(nstate,f+1,x1,a[i]);
            f[0] = x[0] + A_FACTOR * p->h;
          }

/*
a4 requires simply adding
*/

          else
          {
            vadd(nstate,f+1,x1,a[i-1]);
            f[0] = x[0] + p->h;
          }

/*
get the function value
*/

          if( (status=mdiffeq(prog,len,(i == 0 ? x:f),f)) != OK)
            return(status);

/*
multiply f[] by h to get final value of a[i]
*/

          vsmul(nstate,a[i],p->h,f);
        }
```

```
/*
calculate a new value for state vector
*/

        vsmul(nstate,a[1],FACTOR,a[1]);
        vsmul(nstate,a[2],FACTOR,a[2]);

        vadd(nstate,a[0],a[0],a[1]);
        vadd(nstate,a[0],a[0],a[2]);
        vadd(nstate,a[0],a[0],a[3]);

        vsmul(nstate,a[0],ratio,a[0]);

        vadd(nstate,x1,x1,a[0]);

/*
calculate a new value for time
*/

        x[0] = x0 + iter * p->h;

/*
write out the result of the integration
*/

        fputv(stdout,len,x,ROW);

    }

    return(OK);
}
/*
end of rkut
*/
```

The constant vectors $\vec{a}_i$ are stored in the matrix **a[][]**, and the rows of **a[][]** are treated like vectors in the function calls to **vsmul()** and **vadd()**. **a[i]** is a **double** pointer to the ith row of the matrix **a[][]**, which will be passed to the functions exactly like a pointer to a one-dimensional **double** vector would. The inner **for** loop calculates the values for the constants on each iteration of the integration. The vector **f[]** serves to hold the interim values for the state variables, (for $i > 0$) and time during the calculation of the constants, as well as to fetch the function vector. The time + state variable vector is passed directly to **mdiffeq()** for $i = 0$. A conditional expression in the parameter list for **mdiffeq()** makes the selection, thus avoiding having to copy **x[]** into **f[]** for the first iteration of the loop. The final calculation of **x[]** is accomplished through several calls on the routines **vsmul()** and **vadd()**, with **a[0]** serving as an accumulator. The numerical constants in the recursion formula are defined at the top before the function definition using the preprocessor **#define** statement, except for the ratio, $\frac{1}{6}$, which is calculated before the loop begins so that it doesn't have to be calculated each time it is needed in the loop.

6.8.1. Exercises

1. Incorporate the variable step size technique discussed in the exercises for Section 6.6 into the Runge-Kutta routine. For the fourth order Runge-Kutta algorithm, $N = 4$.

2. There are a number of other integration techniques which have been used for integrating differential equations. One group, the predictor-corrector methods, stores values for $\vec{f}$ at preceding iterations and uses them to find the value of $\vec{x}$ on the current iteration. Investigate one of these algorithms and write an integration tool to implement it.

6.9. What Can Go Wrong

The Euler and Runge-Kutta algorithms are useful for solving most initial value problems; however, there is one class of problems with which they have difficulty. An example is the system:

$$\frac{dx_1}{dt} = x_2$$

$$\frac{dx_2}{dt} = 3000x_2 - 3000x_1$$

Although this system is linear, the Runge-Kutta method cannot be used to integrate it.

The problem lies with the large size of the constants in the second equation. If the time constants in one equation are several orders of magnitude larger than those in the others, the time scale of the system will become so distorted that Runge-Kutta methods, or in fact any method with a fixed step size, cannot be used to solve it. Such problems are called stiff problems, and occur in many applications. The warning in Chapter 2 about trying to use Gaussian elimination on matrices with elements several orders of magnitude apart in size stems from similar considerations. Special techniques are required to handle stiff systems, and these are discussed in some of the chapter references.

6.10. Some Examples

In this section, we'll look at two examples of code for model modules. The first example, called **fint.c**, is a model module for the one variable linear differential equation used to produce the data in Tables 6.1 and 6.2. The second example, called **vanderpol.c**, is an implementation of the van der Pol nonlinear oscillator. It is one of the simplest nonlinear differential equations which can exhibit oscillations, and was originally proposed as a model for a vacuum tube oscillator. The dynamic behavior of the van der Pol oscillator is illustrated with a number of graphs, drawn using **graphit**.

Since there are no parameters in the simple linear equation used to generate Tables 6.1 and 6.2, we could have coded **mpgetargs()** to simply do nothing and return. However, as an example of how **mpgetargs()** might work, a rate constant is passed after the **-m** argument:

```
/*
*********************************
mpgetargs-get parameter for equation.
*********************************
*/

int mpgetargs(prog,argch,argbuf)

  char argch;
  char *prog,*argbuf;

{
  extern double rho;
  int sscanf();

    if( sscanf(argbuf,"%lf",&rho) != 1)
    {
      fprintf(stderr,
          "%s:model parameter incorrect.\n",
          prog
          );
      return(ERR);
    }

    return(OK);
}
/*
end of mpgetargs
*/
```

rho is a **static double**, which is used in **mdiffeq()** as the equation parameter. Obviously, a more complicated model may require more complicated parameters, and these could either be passed on the command line or put into a special file, and the name of the file could be passed on the command line.

mdiffeq() for **fint.c** is especially simple. It need only copy the state variable in **x[1]** into the function value in **f[0]**, multiplying by the parameter **rho** in the process:

```
/*
**************************************
mdiffeq-model function for dx/dt = rho * x.
**************************************
*/

mdiffeq(prog,len,x,f)

  int len;
  double x[],f[];
  char *prog;

{
  extern double rho;

/*
check if the number of time + state variable
  vector elements is correct
*/

    if( len > 2 )
      fprintf(stderr,
          "%s:using the first element of input vector.\n",
          prog
          );

/*
now calculate the function
*/


    f[0] = rho * x[1];

    return(OK);

}
/*
end of mdiffeq
*/
```

A warning message is issued if the number of elements in the time + state variable vector exceeds two, and only the first two elements are used. **mdiffeq()** doesn't need to check for fewer than 2 elements, since the main module does that.

A less trivial example is the van der Pol or Lienard system. In vector form, the system is:

$$\frac{dx_1}{dt} = -x_1^3 + \rho x_1 + x_2$$

$$\frac{dx_2}{dt} = x_1$$

For $\rho > 0$, the equations exhibit a periodic orbit in the (x_1, x_2) state plane which attracts all orbits around it. Such an orbit is called a limit cycle, and results in the values of the state variables changing in a periodic way over time. For $\rho < 0$, the point $(0, 0)$ attracts all solutions. In Fig. 6.2, a graph of x_1 against x_2 for $\rho = 1$ shows the cyclic nature of this long-term or steady state solution.

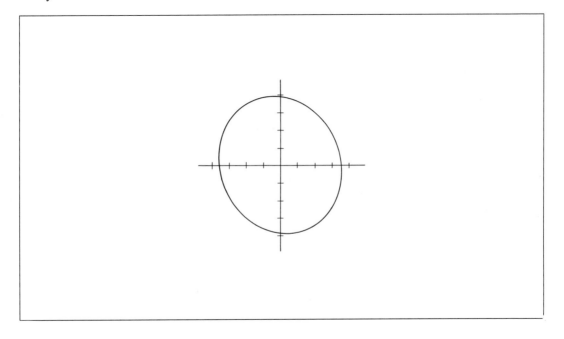

Figure 6.2. A Plot of x_1 vs. x_2 for the van der Pol Equation, $\rho = 1$

The equations were started at the point $(0.1, 0.1)$ for $t_0 = 0.0$ and integrated to $t_m = 50.0$, with a step size of $h = 0.1$. Fig. 6.3 plots x_1 against t, showing that the solutions are periodic. In Fig. 6.4, x_1 is plotted against x_2 for $\rho = -1$. The solution in this parameter range degenerates to the point $(0, 0)$. The integration in Fig. 6.4 was started at $(0.1, 0.1)$ for $t_0 = 0.0$ and integrated to $t_m = 10.0$ with a step size of $h = 0.01$. The plots were generated by using a version of **graphit** linked to a function library for a pen plotter.

The routine for calculating the van der Pol model is a straightforward coding of the equation system:

```
/*
****************************************
mdiffeq-model functions for the Van der Pol
      oscillator.
****************************************
*/

int mdiffeq(prog,len,x,f)

  int len;
  double x[],f[];
  char *prog;

{
  extern double rho;
  double pow();

/*
check if the number of time + state variable
  vector elements is correct
*/

    if( len < 3 )
    {
      fprintf(stderr,
          "%s:not enough time + state variable elements.\n",
          prog
          );
      return(ERR);
    }

    else if( len > 3 )
      fprintf(stderr,
          "vanderpol:using the first 3 elements of input vector.\n"
          );

/*
now calculate the function
*/

    f[0] = -pow(x[1],3.0) + rho * x[1] - x[2];
    f[1] = x[1];
```

```
        return(OK);
}
/*
end of mdiffeq
*/
```

A warning message is issued if too many vector elements are received for the input vector, but the correct number of elements are used anyway. If too few elements are detected, the function returns **ERR**, since calculations cannot proceed.

6.10.1. Exercises

1. Find a differential equation model of a real-world system in the literature and write a model module for it.

2. Modify **main.c**, **euler.c**, and **rkut.c** so that several model functions can be linked with a numerical tool, and the specific model to use can be selected with a command line argument, like the algorithm selection.

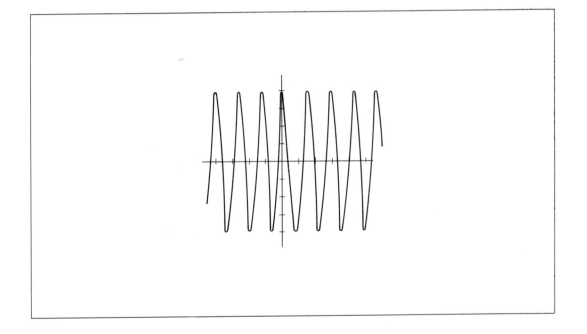

Figure 6.3. A Plot of x_1 vs. t for the van der Pol Equation, $\rho = 1$

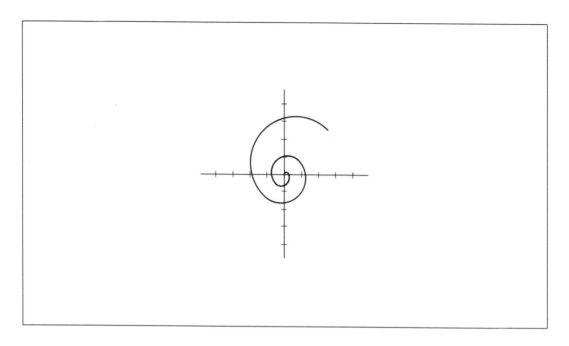

Figure 6.4. A Plot of x_1 vs. x_2 for the van der Pol Equation, $\rho = -1$

6.11. Desertification Modelling

The integration tools in this chapter can be used together with the optimization tools of Chapter 5 to solve problems which combine aspects of both optimization and integration. The following example illustrates:

*   *   *

The spread of the desert into semiarid grazing land, a process called desertification, is a serious problem in many parts of the world, particularly in sub-Saharan Africa. The process is not well understood, but a number of factors seem to be responsible. Weather and geography certainly play a big role, but land use by indigenous peoples, in particular overgrazing, seems to be important as well.

While a definitive model remains to be developed, we can postulate a hypothetical model which shows some of the general properties of a semiarid grazing ecosystem. Typically the grass and other vegetation tend to be under tremendous pressure to reduce water loss, and water availability tends to be the controlling factor in vegetation growth. The vegetation may be thought of as minimizing their potential for water loss at any point in time, by reducing their biomass through dieback if necessary, (during dry periods), and increasing it during wet periods.

This type of model suggests the optimization scenario of Chapter 5. In fact, we can postulate a simple objective function, $V(x, a)$ of plant biomass, x, and grazing animal biomass, a, which models the tendency of plants to minimize their water loss:

$$V(x, a) = \frac{1}{2}x^2 + ax - \alpha x$$

where α is an empirically determined constant.

The water loss potential can be roughly justified as follows. Since water loss is dependent on surface area, a general squared relationship between biomass and surface area might be expected, hence the first term. The second term stems directly from grazing by cattle, while the final term could be expected from the tendency of plants in semiarid areas to reduce their biomass during dry periods by selective dieback. A plant which loses biomass by dieback would reduce its water loss potential, which is why the sign on the last term is negative.

In addition to the model for plant biomass, we need a model for animal growth. We'll use a fairly standard differential equation model from mathematical ecology called the logistic equation. It postulates that growth is proportional to the body weight of the animal and metabolism is proportional to the square of the body weight. Since the animal's growth is dependent on plant material, we'll use x as the constant of proportionality in the growth term:

$$\frac{da}{dt} = a(x - \rho a)$$

where ρ is the proportionality constant for metabolism.

We'll also postulate that the vegetation biomass suffers a large dieback (essentially to zero) when the cattle biomass exceeds β and recovers again when the cattle biomass has been reduced to γ. This may be due to the peripheral effects of cattle grazing, like trampling, etc. During the dieback period, the cattle have nothing to eat, and so begin to starve.

$$* \qquad * \qquad *$$

Note that, in the objective function for water loss potential, a plays the role of a parameter, and likewise, in the animal growth differential equation, x is the growth rate parameter. This model is quite different from the usual differential equation model, since one of the parameters in the system (namely x) will tend to vary, rather than being constant. For this reason, we need to augment the differential equation system with an equation for x, giving:

$$\frac{dx}{dt} = 0$$

$$\frac{da}{dt} = a(x - \rho a)$$

This system basically says that x will remain constant during the integration.

By using the tools developed in Chapter 5 and this chapter, we can write a model module to implement this example. The code for **mdiffeq** needs only call **smethod()** to solve for the value of *x* on each step of the integration, as in the following:

```
/*
***************************************
mdiffeq-model function for grazing example.
***************************************
*/

mdiffeq(prog,len,x,f)

  int len;
  double x[],f[];
  char *prog;

{
  extern struct param p;
  static int on = YES;
  int smethod();

/*
check if the number of time + state variable
  vector elements is correct
*/

    if( len > 3)
      fprintf(stderr,
            "%s:using the first element of input vector.\n",
            prog
            );

    else if( len < 3)
    {
      fprintf(stderr,
            "%s:not enough state variables.\n",
            prog
            );
      return(ERR);
    }

/*
use optimization method to find the value of x
*/
```

```
          if( x[2] >= p.beta || (on == NO && x[2] >= p.gamma) )
          {
            on = NO;
            x[1] = 0;
          }
          else
          {
            on = YES;
            if( smethod(prog,x+1) == ERR)
              return(ERR);
          }

  /*
  now calculate value of function
  */

          f[0] = 0;
          f[1] = x[2] * ( x[1] - 0.1 * x[2] );

          return(OK);
  }
  /*
  end of mdiffeq
  */
```

The parameters α, β, and γ are stored in a **static struct**, **p**. In this example, ρ has been fixed at 0.1. The **struct p** is initialized through the command line argument parsing code in **mpgetargs**(). The time + state variable vector has x in **x[1]** and a in **x[2]**. The function values for the right-hand side are calculated at the end of the routine, after the optimization.

To implement the switching behavior when the limit points of β and γ are reached, we've used a local **static int** called **on**. The system starts with the grass in good condition and **on** is therefore set to **YES**. When **x[2]** exceeds the limit of **p.beta**, then **on** is set to **NO**, since a die-back has occurred. The grass biomass will remain at zero until the cattle biomass is reduced below **p.gamma**, at which point **on** is set to **YES** again and the objective function becomes valid.

The code for the optimization model is equally simple:

```
/*
*************************************
model-grazing ecosystem objective function.
*************************************
*/

int model(x,dervec)

  double x[],dervec[];

{
  extern struct param p;

    dervec[0] = -x[0]  - (x[1] - p.alpha);

    dervec[1] = -1.0;

    return(OK);
}
/*
end of model
*/
```

The optimization algorithm used in the example was Newton's method, discussed in the exercises following Section 5.17, hence, equations for the first and second derivatives are required. Note that, in **mdiffeq()**, **smethod()** is called with **x+1** for the variable vector. This will pass down x as the first vector element in **smethod()** in exactly the right place to be treated as a variable in the optimization. Similarly, **model()** can get at a to calculate the objective function, since it will be in the second vector element. This scheme wouldn't have worked if the dichotomous search method had been used, since both input vector positions are used in **smethod()** for dichotomous search.

Of course, an appropriate **mpgetargs()** and **mpselect()** are needed to set up the parameters for the two model equations. In addition, the parameters for the optimization method must be passed through **mpgetargs()** to **spselect()**. This could be arranged through special flags after the **-m** to indicate which parameter was being passed, or perhaps a file name could be used as an argument, and the parameters for the optimization algorithm and objective function could be read from there.

The results of running the example with $\alpha = 3.0$, $\beta = 1.0$, and $\gamma = 0.5$ are shown in Figure 6.5. The model was started at $t_0 = 0.0$ and $(x,a) = (1.0,0.1)$. The figure plots plant biomass on the y axis against cattle biomass on the x axis, and shows an oscillation, similar to that of the van der Pol model. The cause of the oscillation in this case is quite different, however. In effect, the dieback and growth points make the system a discontinuous one, and force the oscillation. In Figure 6.6, the plant biomass values are plotted against time, again showing the oscillation. The integration was done using the Runge-Kutta method, with **p.iter** = 2000 and **p.h** = 0.01. The parameters for Newton's method were **sp.eps** = 0.01 and **sp.iter** = 25. In all cases, the numerical optimization converged within the limit on the number of iterations.

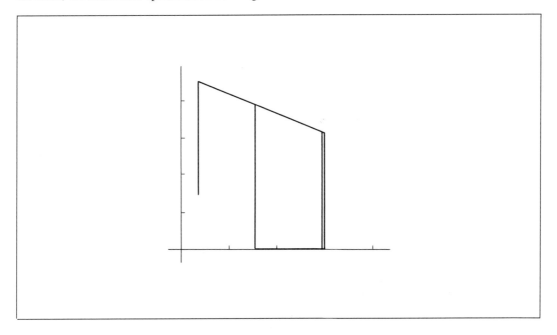

Figure 6.5. Plant Biomass vs. Cattle Biomass

A call graph for this example may be helpful for illustrating how the optimization and integration tool modules fit together:

> **I:main()**
> **>>I:getargs()**
> **>>>>I:mpgetargs()**
> **>>>>>>O:spselect()**
> **>>>>>>O:mpselect()**
> **>>I:euler()**
> **>>I:rkut()**
> **>>>>I:mdiffeq()**
> **>>>>>>O:smethod()**
> **>>>>>>>>O:model()**

The routines which belong to the integration tool modules are prefixed with **I:** while those which are part of the optimization package are prefixed with **O:**. If x had been a vector instead of a scalar, the vector optimization tools could have been used instead.

A disclaimer: this model is not meant to be scientifically accurate, but rather to illustrate how the optimization tools presented in Chapter 5 and the integration tools presented in this

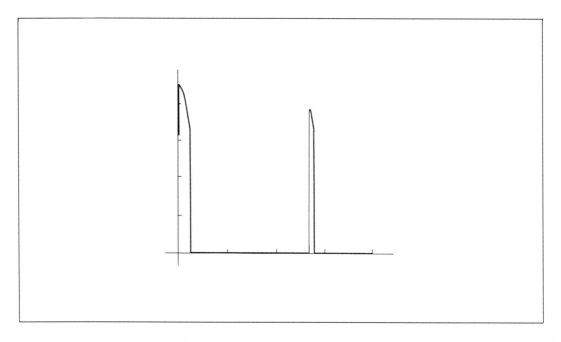

Figure 6.6. Plant Biomass vs. Time

chapter could work together to solve a nontraditional but potentially realistic real-world type model. Although this model is unusual from the numerical integration standpoint (since it combines aspects of both optimization and integration), it may be more realistic for certain real-world systems like the desertification example, which show characteristics of both optimization and integration. Coding the model module for this example took a little over an hour, since most of the work for graphics, numerical integration, and numerical optimization was already completed. This illustrates the advantage of numerical tools: they give you leverage for solving larger problems through combination, since you don't have to rewrite the parts you already have.

6.12. References

Differential equations have been used by mathematical physicists since the time of Newton, so there are many references on the theoretical aspects of differential equations. Most numerical analysis texts also have a chapter or two on numerical methods for solving differential equations. Some good starting points are:

1. *Ordinary Differential Equations with Applications*, by Edward Reiss, Andrew Callegari, and Daljit Ahluwalia, Holt, Rinehart, and Winston, New York, NY, 400 pp., 1976.

    An introduction to the theoretical aspects of differential equations, particularly linear ones.

2. *Differential Equations, Dynamical Systems, and Linear Algebra*, by Morris W. Hirsch and Stephen Smale, Academic Press, New York, NY, 368 pp., 1974.

A more advanced theoretical treatment of differential equations. Contains the van der Pol oscillator example and the logistic equation.

3. *Advanced Calculus for Applications*, by Francis B. Hildebrand, Prentice-Hall, Englewood Cliffs, NJ, 784 pp., 1976.

Discusses deriving Runge-Kutta and other methods from Taylor series expansions.

4. *Numerical Methods*, by Germund Dahlquist, Ake Bjoerck, and Ned Anderson, Prentice-Hall, Englewood Cliffs, NJ, 576 pp., 1974.

Contains discussions of the Euler, Runge-Kutta, and other methods from the numerical perspective, including tips for reducing error.

While the desertification model discussed in the last section is somewhat speculative, models involving differential equations and potential functions have some applicability to other fields, thermodynamics for example. If you're interested in some theoretical background on a different approach to nonlinear optimization and differential equations, try:

1. *Catastrophe Theory and Its Applications*, by Tim Poston and Ian Stewart, Pitman Publishing Ltd., London, 496 pp., 1978.

Index